Integrated High-V_{in} Multi-MHz Converters

Jürgen Wittmann

Integrated High-V_{in} Multi-MHz Converters

 Springer

Jürgen Wittmann
Gröbenzell
Bayern, Germany

ISBN 978-3-030-25259-5   ISBN 978-3-030-25257-1   (eBook)
https://doi.org/10.1007/978-3-030-25257-1

This Springer imprint is published by the registered company Springer Nature Switzerland AG.
The registered company address is: Gewerbestrasse 11, 6330 Cham, Switzerland

To my beloved family, Heloisa and Sebastian, and to my parents, Anna and Franz.

Preface

The work for this book was initially driven by automotive applications, with the goal to do research on how point-of-load voltage converters, supplying electronic components, can be made smaller and cheaper in order to reduce size and cost of the electronics. The main objective was to investigate the impact of an increasing switching frequency into the multi-MHz range to reduce the size of the passives in inductive switching converters. The initial target was to supply the converters directly from a conventional 12 V car battery. As automotive applications need to tolerate largely varying battery voltages up to 40 V and above, it was set by the applications that the increase of the switching frequency needs to be investigated in combination with high input voltages of the converters. This combination represents a main limitation not only for the converter efficiency but also for the circuit design, which has been the main driver for this work. During the research of the work for this book, a trend emerged in several applications to increase the system supply voltage. Higher supply voltages enable a more efficient power delivery to the electronic components with lower currents and thus lower wiring costs. In consequence, the maximum power provided to the system can be increased. To mention some examples, the automotive industry introduced a standard for a new supply domain with a 48 V battery, as the conventional 12 V board net was at its power limit due to new functionalities, e.g., autonomous driving. E-mobility applications enter the market with a wide range of battery voltages up to 60 V and above, which required to cope with the large power demand for driving. In the field of IT, server supply concepts were redefined to increase the rack level voltage distribution from typically 12 to 48 V to reduce power losses and the energy costs in large data centers. The latest USB Power Delivery standard increased the charging voltage up to 20 V to allow fast-charging of mobile and IoT devices. The demand for compact and highly integrated voltage converters, suitable for increasing input voltages, arose from several different areas, which finally set the focus of the work for this book.

Many publications for highly integrated switching converters can be found for low input voltages (typ. <5 V). Publications on converters for higher input voltages ($\gg$5 V) show a low integration level with large and bulky components and make systems costly. Highly integrated, fast-switching converters for higher input voltages, as high as 50 V, are not or only rarely covered.

The purpose of this book is to provide insight into the trade-offs, which are required to achieve an acceptable efficiency with both a high integration (small size), mainly addressed with faster switching frequencies, and high input voltages. Achieving switching frequencies in the order of 10–25 MHz at high input voltages brings the converters to its limit in terms of circuit design, implementation, and technology. This book describes system-level aspects and circuit design techniques to push out and to overcome these limits. A comparison of converter types for different operating ranges is presented along with practical implementation examples. Soft-switching architectures are studied in comparison to conventional buck converter architectures. The purpose of this book is to give the reader a guideline for selecting the right converter architecture and to point out potential issues and limitations related to the chosen architecture. It covers converter theory, architectures, circuit design, efficiency, sizing of passives, PCB design, and technology aspects. Besides the focus on fast-switching and high input voltages, the contents of the book are applicable to any type of switching converter implementation.

The research project resulting in this book was sponsored by Robert Bosch GmbH, Reutlingen, Germany. The project was executed at Reutlingen University, Germany, in cooperation with the University of Stuttgart, Germany, and the Leibniz University Hannover, Germany. I would like to express my sincere gratitude to my advisor, Prof. Bernhard Wicht, now head of Mixed-Signal Circuits Group, Institute of Microelectronic Systems at Leibniz University Hannover. He was at my side as a mentor along most of my engineer career. His continuous support and trust were the bases for my technical growth and finally for this book. I enjoyed our long-going discussions resulting in fruitful ideas, and I highly appreciated his support and contribution, also in structuring, writing, and reviewing this book. I want to thank Dr. Axel Wenzler, Thoralf Rosahl, and Stefen Ritzmann of Robert Bosch GmbH for their valuable industry inputs and reviews and all other colleagues at Bosch, who supported this work with various test chips. I wish to thank Prof. Jörg Schulze, University of Stuttgart, and his team for providing me valuable technology support, as well as for being part of his research seminars and his useful cross-disciplinary feedback, which helped me to structure this work. I was very pleased that I had Prof. Dongsheng Brian Ma, University of Texas, Dallas, USA, as a reviewer of this work, to receive his valuable feedback based on his profound power management expertise. I want to say thank you to all my colleagues and friends at Reutlingen University, especially to Tobias Funk, Achim Seidl, Alexander Barner, and Christoph Rindfleisch, who supported this work with their master's theses.

Finally, I want to express my deepest gratitude to my loving wife, Heloisa, and to my wonderful son, Sebastian. I want to thank you so much for always being at my side and giving me all the emotional support and motivation to go on with writing this book. Without your understanding, your patience, and your great support in organizing our life, it would not have been possible to complete this book.

Gröbenzell, Germany   Jürgen Wittmann
May 2019

Contents

1 Introduction 1
  1.1 Scope of This Book 3
  1.2 Outline of This Book 4
      1.2.1 Motivation for High-V_{in} Converters and Fundamentals 4
      1.2.2 Fast-Switching High-V_{in} Buck Converters 4
      1.2.3 Design of Fast-Switching Circuit Blocks 5
      1.2.4 Efficiency and Loss Modeling of High-V_{in} Multi-MHz
              Converters 5
      1.2.5 Dead Time Control 6
      1.2.6 Resonant Converters 6
      1.2.7 Conclusion and Outlook 7
  References 7

2 Motivation for High-V_{in} Converters and Fundamentals 9
  2.1 Applications for High-V_{in} DC-DC Converters 9
      2.1.1 Automotive and e-Mobility 9
      2.1.2 Servers 12
      2.1.3 Automation 15
      2.1.4 Mobile Devices 16
  2.2 Power Conversion to Low-Voltage Components 16
      2.2.1 DC-DC Conversion Fundamentals 16
      2.2.2 Types of DC-DC Converters 18
  2.3 Buck Converter Fundamentals 26
      2.3.1 Buck Converter Parameters 26
      2.3.2 Scaling of the Output Filter 28
      2.3.3 Scaling of Real Inductors 30
      2.3.4 Scaling of Real Capacitors 33
      2.3.5 Losses in a Buck Converter 34
      2.3.6 Design Challenges for High-V_{in} 39
  2.4 State-of-the-Art Converters 41
  Appendix 42
  References 43

3 Fast-Switching High-V_{in} Buck Converters 47
   3.1 Buck Converter Architectures 47
   3.2 Power Switch Technologies 51
   3.3 Converter Implementation Aspects 53
   3.4 Substrate Coupling 56
      3.4.1 Comparison of Isolation Structures by Simulation 58
      3.4.2 Verification of Isolation Structures by Measurements 62
   References 64

4 Design of Fast-Switching Circuit Blocks 67
   4.1 Output Voltage Regulation 67
   4.2 Sawtooth and PWM Generator 69
      4.2.1 Requirements for Sawtooth and PWM Generator 69
      4.2.2 Limitations of Conventional Sawtooth Generators 70
   4.3 Level Shifters 75
      4.3.1 Conventional Level Shifter Concepts 76
      4.3.2 Level Shifter for PMOS High-Side Switches 77
      4.3.3 Level Shifter for NMOS High-Side Switches 78
   4.4 Gate Driver 81
   4.5 Experimental Results of Fast-Switching Circuit Blocks 83
   References 86

5 Efficiency and Loss Modeling of High-V_{in} Multi-MHz Converters 89
   5.1 Conventional Efficiency Modeling 90
   5.2 Priorities of Loss Modeling for High-V_{in} Fast-Switching Converters   90
   5.3 Efficiency Model for an Asynchronous Buck Converter 92
      5.3.1 Diode Conduction Losses 93
      5.3.2 Switching Behavior 93
      5.3.3 Implementation as Four-Phase Model 96
      5.3.4 Root Cause and Loss Location Analysis 98
      5.3.5 Contribution and Modeling of Non-linear Capacitances 99
   5.4 Loss Contributions in a Synchronous Buck Converter 104
      5.4.1 Contribution of the Low-Side Switch 104
      5.4.2 Influence of Dead Time on Switching Behavior 105
      5.4.3 Dead Time Related Losses 106
      5.4.4 Dead Time Related Losses at Light Load 110
   5.5 Loss Optimization and Limitations 113
   5.6 Architecture Comparison 114
   5.7 Design Indicator: Efficiency Scaling 117
   Appendix 120
   References 122

6 Dead Time Control 125
   6.1 State-of-the-Art Dead Time Controls 125
   6.2 Predictive Mixed-Signal Dead Time Control 126
      6.2.1 Dead Time Control Concept 126

		6.2.2	Sample and Hold Circuit	128
		6.2.3	Differential Delay Lines	130
		6.2.4	Experimental Results	132
	6.3	Enhanced Dead Time Control for Light Load		136
		6.3.1	Turn-High Dead Time Control Concept	137
		6.3.2	Experimental Results	138
	References			139

7 Resonant Converters ... 141
7.1 Quasi-Resonant Converter ... 141
7.2 Parallel-Resonant Converter (PRC) ... 144
    7.2.1 Switching Concept and Operating Modes ... 145
    7.2.2 Converter Implementation ... 147
    7.2.3 Experimental Results ... 149
References ... 157

8 Conclusion and Outlook ... 161
8.1 Conclusion ... 161
8.2 Comparison to State-of-the-Art ... 164
8.3 Outlook ... 165
References ... 167

Index ... 169

Acronyms

List of Abbreviations

ASIC	Application-specific integrated circuit
BSM	Back-side metallization
CCM	Continuous conduction mode
CMC	Current mode control
CTO	Chief technical officer
DCM	Discontinuous conduction mode
DMOS	Double-diffused metal-oxide semiconductor
ECU	Electronic control unit
EPI	Epitaxial layer
ESR	Equivalent series resistance
ESV	Electric sports vehicle
EUV	Electric utility vehicle
FET	Field-effect transistor
GaN	Gallium nitride
IC	Integrated circuit
IGBT	Insulated gate bipolar transistor
IoT	Internet of Things
LDMOS	Laterally diffused metal-oxide semiconductor
LEV	Light-electric vehicle
MOSFET	Metal-oxide-semiconductor field-effect transistor
NMOS	n-Channel metal-oxide semiconductor
PCB	Printed circuit board
PDU	Power distribution unit
PMOS	p-Channel metal-oxide semiconductor
PRC	Parallel-resonant converter
PSU	Power supply unit
PUE	Power usage effectiveness
PWM	Pulse-width modulation

QRC	Quasi-resonant converter
SC	Switched capacitor
SELV	Safety extra-low voltage
SI	Silicon
SIP	System in package
SMD	Surface-mount device
SMPS	Switched-mode power supply
SoC	Systems on a chip
SoI	Silicon on insulator
SPUE	Server PUE
TCAD	Technology computer-aided design
UPS	Uninterruptable power supply
VMC	Voltage-mode control
VR	Voltage regulator
WLCSP	Wafer-level chip scale package
ZCD	Zero-cross detection
ZVS	Zero-voltage switching

List of Symbols

α		Factor of increasing driver strength in a tapered gate driver
C_0	F	Capacitor of an output filter in a buck converter
C_1	F	Capacitor of the frequency compensated voltage divider in the S&H
C_2	F	Capacitor of the frequency compensated voltage divider in the S&H
C_{vdd5}	F	Buffer capacitor of the low-side supply
C_{boot}	F	Bootstrap buffer capacitor
C_c	F	Capacitance of an ideal capacitor
C_{db}	F	Drain-bulk capacitance of a power transistor
C_{del}	F	Delay capacitance in the differential delay element
C_{dio}	F	Reverse blocking capacitance of a diode
C_{ds}	F	Capacitance at the drain of the transistor
$C_{fly,n}$	F	Flying capacitors
C_g	F	Gate capacitance of a power transistor
C_{gd}	F	Gate-drain capacitance of a power transistor
$C_{gd,in}$	F	Input capacitance at the gate driver
C_{gs}	F	Gate-source capacitance of a power transistor
C_h	C	Hold capacitor of the S&H
C_{HSGND}	F	Buffer capacitor at the high-side supply of a PMOS switch
C_{in}	F	Buffer capacitance at the converter input
$CLK1$	V	CLK1 signal of the first PWM generator stage
$CLK2$	V	CLK2 signal of the second PWM generator stage

$C_{p,ls}$	F	Parasitic capacitance from the high side to the signal path of the level shifter
C_r	F	Capacitor of resonant circuit in a resonant converter
C_s	F	Sample capacitor of the S&H
C_{sub}	F	Parasitic substrate capacitance
C_{sw}	F	Capacitance of the switching node
$CTRL_{hs}$	V	Control signal of the high-side switch
$CTRL_{ls}$	V	Control signal of the low-side switch
D		Duty cycle of a pulse-width modulated signal
D_0		Freewheeling diode in a resonant converter
D_b		Body diode of the power switch
$\Delta\eta$	%	Improvement of power efficiency
DH_A		Diode clamp connected to V_{boot} in the NMOS level shifter
DH_B		Diode clamp connected to V_{boot} in the NMOS level shifter
DI		Design Indicator
Δi_{L0}	A	Current in L_0
D_L		Freewheeling diode in the asynchronous buck converter
ΔI_{sig}	A	Differential signal current in the NMOS level shifter
DL_A		Diode clamp connected to V_{sw} in the NMOS level shifter
DL_B		Diode clamp connected to V_{sw} in the NMOS level shifter
D_{out}	V	Non-delayed output signal of the delay line
$D_{out,del}$	V	Delayed output signal of the delay line
ΔP_{losses}	%	Improvements of switching losses
DT_{hi}	s	Dead time at the converter's turn-high transition
DT_{lo}	s	Dead time at the converter's turn-low transition
Δv_{out}	V	Output voltage ripple of a converter
ΔV_{sig}	V	Differential voltage at the comparator input in the NMOS level shifter
η		Overall efficiency of a DC-DC converters
η_1		Efficiency of a first cascaded DC-DC converter
η_2		Efficiency of a second cascaded DC-DC converter
η_{ls}		Efficiency of the low-side gate driver supply generation
η_{lin}		Efficiency of a linear regulator
η_{opt}		Optimal power efficiency
η_{SC}		Efficiency of a switched-capacitor converter
f_{sw}	Hz	Open-loop crossover frequency of a voltage mode control
f_{sw}	Hz	Frequency of a switching converter (switching frequency)
f_{sw0}	Hz	Initial frequency of a switching converter (switching frequency)
GND	V	System ground
$HSGND$	V	High-side ground
I_1	A	Intermediate current between cascaded DC and DC converters
I_{C0}	A	Current into the buck converter capacitor C_0
$I_{coupling}$	A	Coupling currents in the NMOS level shifter
$I_{cpl,sig}$	A	Coupling current in the level shifter
I_d	A	Drain current of the power switch

I_{dio}	A	Diode forward current
I_{in}	A	Input current
I_{L0}	A	Current in L_0
I_{Lr}	A	Current through the resonant inductor
I_{Lmax}	A	Maximum inductor current of a buck converter
I_{Lmin}	A	Minimum inductor current of a buck converter
I_m	A	Motor supply current
I_{out}	A	Output current
I_{out0}	A	Initial output current
$I_{out,max}$	A	Maximum output current
$I_{out,tr0}$	A	Reference load current parameter for rising voltage transition calculation
I_{pu}	A	Pull-up current to to achieve soft-switching at the high-side switch turn-on
I_{ramp}	A	Constant current generating the sawtooth ramp
I_{ramp1}	A	Constant current generating the sawtooth ramp
I_{ramp2}	A	Constant current generating the sawtooth ramp
$I_{cpl,sig}$	A	Signal current in the level shifter
I_{sub}	A	Coupling currents into the substrate
I_{up}	A	Current source load in the NMOS level shifter
L_0	H	Inductor of an output filter in a buck converter
L_{boot}	H	Parasitic inductance at the converter's boot supply input
$L_{gnd,vdd5}$	H	Parasitic inductance at the converter's low-side supply ground input
L_{in}	H	Parasitic inductance at the converter's input
L_m	H	Motor inductance
$L_{lp,gnd}$	H	Parasitic inductance at the converter's power ground input
L_r	H	Inductor of resonant circuit in a resonant converter
L_{sw}	H	Parasitic inductance at the converter's switching node
$L_{sw,boot}$	H	Parasitic inductance at the converter's boot supply reference input
L_{vdd5}	H	Parasitic inductance at the converter's low-side supply input
MN_0		Low-side NMOS switch of a level shifter
MN0		Main switch of the PRC
MN_1		Low-side NMOS switch of a level shifter
MN_{clp}		Clamping transistor in the NMOS level shifter
MN_{HS}		High-side NMOS switch
MN_{LS}		Low-side NMOS switch
MNR		Resonant switch in the PRC
MP_1		High-side current mirror input of the PMOS level shifter
MP_2		High-side current mirror input of the PMOS level shifter
MP_{casc}		High-side current mirror input of the PMOS level shifter
MP_{clp}		Clamping transistor in the NMOS level shifter
MP_{hs}		High-side PMOS switch
MP_{lin}		Regulation transistor of the linear regulator generating $HSGND$
MP_{ls}		Low-side PMOS switch

MPR		PMOS power switch in the resonant circuit of the PRC
N_i		Ideal conversion ratio
P_1	W	Intermediate power between cascaded DC and DC converters
P_{avg}	W	Average losses
P_{chg}	W	Charging losses of the switching node at the high-side switch turn-on
P_{cond}	W	Conduction losses of a buck converter
$P_{cond,hs}$	W	Conduction losses of the high-side power switch
$P_{cond,ls}$	W	Conduction losses of the low-side power switch
P_{DThi}	W	Losses caused by the dead time at the converter's turn-high transition
P_{DTlo}	W	Losses caused by the dead time at the converter's turn-high transition
φ_1	V	Switching phase 1 of a switched-capacitor converter
φ_2	V	Switching phase 2 of a switched-capacitor converter
$\overline{\Phi}$	V	Inverted switch control signal of the sampling stage
$\overline{\Phi_{out}}$	V	Inverted switch control signal of the hold stage
Φ_{out}	V	Switch control signal of the hold stage
Φ	V	Switch control signal of the sampling stage
P_{in}	W	Converter's input power
$P_{L,DC}$	W	DC losses of an inductor
P_{loss}	W	Power losses in a converter
P_{loss0}	W	Power losses in an initial operating point
$P_{loss,norm}$		Normalized losses
P_{out}	W	Converter's output power
P_{out0}	W	Initial output power of a converter
P_{tr}	W	Transition losses in a buck converter
PWM	V	Pulse-width modulated signal
PWM_0	V	PWM signal at the low-side to control the main switch in the PRC
$PWM1$	V	PWM signal of the first PWM generator stage
$PWM2$	V	PWM signal of the second PWM generator stage
$CTRL_{hs}$	V	High-side switch control signal generated by the dead time controller
$CTRL_{ls}$	V	Low-side switch control signal generated by the dead time controller
PWM_{MN0}	V	PWM signal to control the main switch in the PRC
$\overline{PWM}$	V	Inverted pulse-width modulated signal
PWM_r	V	PWM signal at the low-side to control the resonant switch in the PRC
Q_1	C	Charge on an ideal capacitor
Q_2	C	Charge on an ideal capacitor
Q_c	C	Charge on an ideal capacitor
Q_{dio}	C	Charge on the reverse blocking capacitance of a diode

Q_{dio1}	C	Charge on the reverse blocking capacitance of a diode at the voltage V_{dio1}
Q_{dio2}	C	Charge on the reverse blocking capacitance of a diode at the voltage V_{src}
Q_g	C	Gate charge required to turn on a MOSFET
$Q_{\text{gs},1}$	C	Charge flowing to C_{gs} when the power switch turns on
$Q_{\text{gd},1}$	C	Charge flowing to C_{gd} when the power switch turns on
Q_{oss}	C	Drain charge flowing in a MOSFET during the on-state switching event
Q_{rr}	C	Reverse recovery charge
R		General resistor
R_1	C	Resistor of the frequency compensated voltage divider in the S&H
R_2	C	Resistor of the frequency compensated voltage divider in the S&H
R_{dcr}	Ω	DC resistance of an inductor
R_{esr}	Ω	DC resistance of a capacitor
R_{lin}	Ω	Resistance of a linear regulator
R_{load}	Ω	Load of a converter
R_{on}	Ω	On-state resistance of a power switch in a converter
$R_{\text{on,hs}}$	Ω	On-state resistance of a high-side power switch
$R_{\text{on,ls}}$	Ω	On-state resistance of a low-side power switch
R_{sc}	Ω	Equivalent resistance
R_{sub}	Ω	Resistance along the substrate
S_1	s	First sampling instant of the S&H stage
S_2	s	Second sampling instant of the dead time control
S_H		High-side switch of a buck converter
S_L		Low-side switch of a buck converter
T	s	Switching period of a switching converter
t_{dio1}	s	Diode conduction time during DT_{hi}
t_{dio2}	s	Diode conduction time during DT_{lo}
$t_{\text{dio,hs}}$	s	Diode conduction time of the high-side switch during DT_{hi} in light load
$t_{\text{d,ps}}$	s	Propagation delay of the power stage
$t_{\text{d,r}}$	s	Delay to turn on the main switch in the resonant converter
$t_{\text{d,sw}}$	s	Delay to turn on the main switch in the resonant converter
t_f	s	Fall time of the switching and resonant node in the quasi-resonant converter
t_{fall}	s	Duration of the falling edge of the sawtooth signal
$t_{\text{ls,off}}$	s	Time at when the high-side switch is turned off
$t_{\text{hs,on}}$	s	Time at when the high-side switch is turned on
$t_{\text{ILr,rise}}$	s	Rise time of the current in the resonant inductor
t	s	Time
$t_{\text{ls,off}}$	s	Time at when the low-side switch is turned off
$t_{\text{ls,on}}$	s	Time at when the low-side switch is turned on

t_{mux}	s	Time required for a multiplexed sawtooth signal to settle
t_{off}	s	Off-time of PWM
t_{on}	s	On-time of PWM
$t_{on,sw}$	s	High time of V_{sw} in the quasi-resonant converter
t_{osc}	s	Oscillation time of resonant node in the quasi-resonant converter
t_{pd}	s	Propagation delay of a comparator
t_r	s	Time of the converter's rising transition
t_{ri}	s	Current transition at the converter's rising transition
t_{rv}	s	Time of voltage transition during the converter's rising transition
t_{rv0}	s	Reference parameter for rising voltage transition calculation
$turn\_on$	V	Switch turn-on signal in the PRC
t_{vi}	s	Time of full VI overlap during the converter's rising transition
$t_{vi0,a}$	s	Load current independent time in the converter's rising transition
$t_{vi0,b}$	s	Load current dependent time in the converter's rising transition
V_1	V	Intermediate voltage between cascaded DC-DC converters
V_{boot}	V	Bootstrap (high-side) supply, voltage at C_{boot}
$V_{boot,i}$	V	Internal bootstrap (high-side) supply
V_c	V	Voltage of an ideal capacitor
VCR		Conversion ratio of a DC-DC converter
VCR_{ideal}	I	Ideal conversion ratio of a switched-capacitor DC-DC converter
V_d	V	Drain voltage of a transistor
V_{dd5}	V	Supply voltage of the 5 V low-voltage domain
V_{dio}	V	Reverse blocking voltage of a diode
V_{dio1}	V	Reverse blocking voltage of a diode at the charge Q_{dio1}
V_{ds}	V	Drain-source voltage of a transistor
$V_{ds,MN0}$	V	Drain-source voltage of the main switch in the PRC
V_{err}	V	Output of the error amplifier
V_f	V	Body diode of the power switch
V_g	V	Gate voltage of a transistor
V_{gs}	V	Gate-source voltage of a transistor
$V_{gs,hs}$	V	Gate-source voltage of the high-side transistor
$V_{gs,ls}$	V	Gate-source voltage of the low-side transistor
V_{hi}	V	Upper reference of the window comparators
V_{in}	V	Input voltage
V_{in0}	V	Initial input voltage
$V_{in,tr0}$	s	Reference input voltage parameter for rising voltage transition calculation
V_{L0}	V	Voltage over L_0
V_{lo}	V	Lower reference of the window comparators
V_m	V	Motor supply voltage
V_n	s	Signal of the last low-side gate driver stage
V_{on}	V	Voltage drop caused by R_{on} of a conducting switch
V_{out}	V	Output voltage
V_{out0}	V	Initial output voltage
$V_{out,ideal}$	V	Ideal output voltage

V_{Q1}	V	Voltage of an ideal capacitor
V_r	V	Resonant node of a resonant converter
$\hat{V}_r$	V	Amplitude of the resonant node of a resonant converter
V_{ramp1}	V	Ramp signal of the PWM generator
V_{ramp2}	V	Ramp signal of the PWM generator
V_{ref}	V	Reference voltage of the frequency compensated voltage divider
$V_{ref,hi}$	V	Upper sawtooth signal reference
$V_{ref,lo}$	V	Lower sawtooth signal reference
V_s	V	Sampled switching node voltage at C_s
V_{sh}	V	Output voltage of the sample and hold stage at C_h
$V_{sh1/2}$	V	Output voltage of the sample and hold stage at C_h
V_{sink}	V	Substrate voltage shift at the current sink
V_{src}	V	General voltage source
V_{sw}	V	Switching node voltage of a buck converter
$V_{sw,i}$	V	Internal switching node voltage of a buck converter
$V_{sw,lv}$	V	Divided switching node voltage V_{sw}
V_{th}	V	Threshold of a transistor
$V_{ton,0}$	V	Reference to adjust the switch on-time of MP0 in the PRC
$V_{ton,r}$	V	Reference to adjust the switch on-time of MPR in the PRC
W_c	J	Stored energy on an ideal capacitor
W_{c1}	J	Stored energy on an ideal capacitor
W_{c2}	J	Stored energy on an ideal capacitor
$W_{DThi,short}$	J	Loss energy caused by a too short dead time DT_{hi} at light load
$W_{loss,chrg}$	J	Energy lost by charging a capacitor
$W_{loss,chrg1}$	J	Energy lost by charging a capacitor
$W_{loss,chrg2}$	J	Energy lost by charging a capacitor
W_{qrr}	J	Eenergy lost due to reverse recovery effect
W_{src}	J	Energy delivered by the voltage source V_{src}
W_{sw}	J	Eenergy lost during the charging event of C_{sw}
ZCD_{Vr}	V	Zero-cross detection signal of V_r in the PRC
ZCD_{Vsw}	V	Zero-cross detection signal of V_{sw} in the PRC

Chapter 1
Introduction

The continuous trend towards higher integration of electrical circuits enables many new functions in a wide range of applications. Systems-on-a-chip (SoC) has become common, in which often most or all of the electrical functionalities are realized on one single chip. The power management is one fundamental key aspect to be considered to achieve a high integration of these systems. Compact voltage converters are required to be integrated within an SoC or close to the point-of-load. A wide range of possible system supply voltages and a high number of different power specifications of the supplied components often require a specific voltage converter for each component in an electrical system.

The trend to more functionality comes along with a higher power demand. This initiated an increase of the system supply voltage to enable a more efficient power distribution. In recent years, a change in the power supply concept could be observed in several applications in areas like e-mobility and automotive, servers, automation, mobile devices, and several others.

Mobility and automotive applications experience a strong push towards driver assistant systems and fully autonomous driving. This requires a further increase of the already high number of existing electrical components in a conventional car, to provide comfort, infotainment, safety, lightning, communication, and many other functions. A second trend is the electrification of mobility in form of hybrid and fully battery driven vehicles, like e-bikes, e-scooters, cars, and trucks. The first fully electric autonomous driving cars are available on the market [14, 17]. A disruptive mobility technology might be drones, which have been shown to be suitable for transportation of persons and goods [3]. The electrification and increasing amount of components lead to a significantly higher power demand. Light electric vehicles (LEV) are designed for battery voltages towards 48 V, or even higher, to increase the available battery power. The conventional 12 V board net in cars is reaching the power limit, with the consequence that an additional 48 V board net was introduced to supply high-power applications [9]. The electrical components have to be supplied by DC–DC converters at the point-of-load, operating from a wide range

© Springer Nature Switzerland AG 2020
J. Wittmann, *Integrated High-V*$_{in}$ *Multi-MHz Converters*,
https://doi.org/10.1007/978-3-030-25257-1_1

of different and varying supply voltages from below 12 V up to 60 V and higher. Higher supply voltages limit the size and efficiency of these DC–DC converters. The increasing amount of components, and thus the number of DC–DC converters, lead to an even higher demand for high integration and cost reduction.

In servers, the trends towards mobile and cloud applications, Internet of Things (IoT), Big Data, and others, lead to fast increasing number of data, which are processed in servers in huge data centers with high processing and storage capability. Due to the large amount of servers worldwide, the server and also the power distribution efficiency has to be increased, as it highly impacts the worldwide energy consumption and the data center's profitability. Low-voltage server components, like CPU, RAM, hard disks, or network interfaces, are conventionally supplied by point-of-load DC–DC converters from a 12 V power supply. On the one hand, the conversion ratio of these point-of-load DC–DC converters is increasing, as the supply voltages of low-voltage components are continuously decreasing, along with improving technology nodes, towards voltages below 1 V. On the other hand, it has been shown that the voltage distribution within a server rack can be more efficient if the 12 V power supply is increased towards 48 V [10, 27].

Several other applications follow this trend towards higher system supply voltages. DC–DC converters have to cope with this trend towards increasing input voltages (V_{in}) and decreasing output voltages (V_{out}), resulting in high conversion ratios. While for low-voltage battery supplies, e.g., in mobile phones, very compact DC–DC converters are available [1, 2, 4–8, 11–13, 15, 16, 18–26], size, cost, and efficiency are strongly limited at higher supply voltages towards 48 V. State-of-the-art converters achieve a high conversion efficiency, but they often dominate the size and cost of the applications, as they are operated at low switching frequencies. The research focus for high-integration and cost reduction in the past years was set mainly on low-voltage DC–DC converters with input voltages limited to 5 V and below, and are poorly covered for higher input voltages.

High-voltage switches and capacitors make switched capacitor converters inefficient. Therefore, inductive switched-mode DC–DC converters are usually preferred at high input voltages. High conversion ratios (V_{in}/V_{out}) and low output voltages (V_{out}), are limited by small duty cycles, resulting in a short power switch on-times in pulse-width modulated (PWM) converters. Moreover, switching high input voltages significantly increases the switching losses and thus reduces the converter's power efficiency. An increase of the switching frequency drives a significant reduction of the converter size and cost, and enables a high level of integration, as the size of filter components scales down. As a drawback, this leads to a further increase of the switching losses. In addition, higher switching frequencies at small duty cycles require a further on-time reduction of the PWM control signal of the power switches, what affects fast-switching circuit blocks of the converter to be implemented in bandwidth limited high-voltage technologies. Decreasing power efficiency and reducing PWM on-times by circuit design represent the major barrier for a further converter size reduction towards a full integration.

1.1 Scope of This Book

The scope of this book is summarized and depicted in Fig. 1.1. It is strongly related to the trend towards increasing system supply voltages, driven by a rising power demand and new features in several applications. Point-of-load voltage converters are required to down-regulate significantly increasing system supply voltages. This book covers converters with integrated power switches in standard silicon technologies delivering an output power of typically up to 10–25 W, with the focus on non-isolated inductive switched-mode converter architectures. Requirements in terms of cost, size, and better reliability are addressed by an improving level of integration. In particular, converters are operated at increasing switching frequencies in order to scale down the size of dominant filter components.

The limitations of point-of-load converters in the reduced minimum PWM on-time, and in the decreasing efficiency, are addressed by three main topics. (1) Implementation aspects of the converter, specifically circuit block design, technology, and layout of the printed circuit board PCB, are investigated. New design concepts and

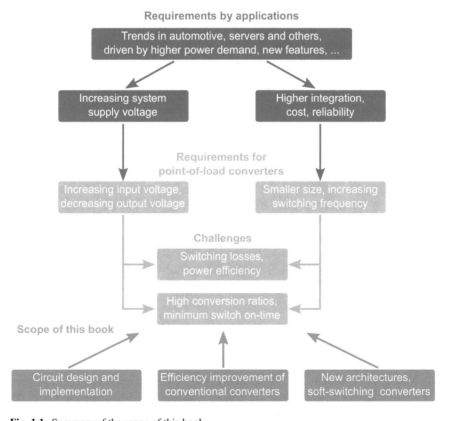

Fig. 1.1 Summary of the scope of this book

circuits are developed to minimize the minimum switch on-time, allowing faster and more robust switching at higher voltages. (2) The efficiency of conventional buck converters is analyzed in order to compare different implementations. The main loss contributors are investigated and addressed to improve efficiency. (3) New architectures and more efficient soft-switching techniques are investigated to allow an even further increase of the switching frequency at acceptable efficiencies.

1.2 Outline of This Book

This section describes the structure of this book, and summarizes the particular chapters and sections.

1.2.1 Motivation for High-V_{in} Converters and Fundamentals

Chapter 2 describes the motivation for this book, and the requirements for high-V_{in} multi-MHz converters. Applications using highly-integrated point-of-load converter with high input voltages are reviewed. The system requirements for voltage converters, especially for automotive and servers, are summarized and compared (Sect. 2.1). System level aspects of a voltage conversion are covered, which demonstrates that a single-step conversion from high input voltages to the point-of-load is most beneficial for size and efficiency (Sect. 2.2). Different DC–DC converter architectures are reviewed, including linear regulators, switched capacitor converters, inductive converters, soft- and resonant converters, and hybrid converters, pointing out the advantages of inductive converters for this type of applications (Sect. 2.2.2). Inductive buck converter fundamentals demonstrate that the switching frequency and current ripple, as well as the output voltage and the conversion ratio, have a main influence on the converter size, and also efficiency. The benefits of soft-switching and resonant converters are introduced. A study of the scaling of commercial filter inductors and capacitors with respect to switching frequency and current ripple is presented, and the limitations of higher switching frequencies at increasing input voltages are discussed (Sect. 2.3). A study of commercially available and published state-of-the-art converters for these applications is shown (Sect. 2.4).

1.2.2 Fast-Switching High-V_{in} Buck Converters

Chapter 3 elaborates the main issues arising with common implementation variants of a buck converter. These issues are addressed further throughout this book. An asynchronous buck converter with a low-side freewheeling diode and a synchronous buck converter with a high-side and low-side switch are compared with respect to high input voltages and multi-MHz switching. An implementation of the architectures with PMOS (p-channel MOSFET) and NMOS (n-channel MOSFET) power switches is evaluated. The impact on efficiency is discussed, and the requirements for the converter and circuit block implementations are derived (Sect. 3.1). Different switch technologies, in particular integrated lateral DMOS (double-diffused metal-oxide semiconductor) transistors, external vertical switches, and gallium nitride (GaN) switches are compared (Sect. 3.2). The influence of parasitics caused by the PCB is discussed, which cause large voltage ringing at the power switches. It is demonstrated that an optimization of the PCB including a direct-bond to PCB technique significantly reduces ringing (Sect. 3.3). The isolation of a switching high-side domain, controlling an NMOS switch, induces substrate coupling, which results in malfunction of the converter. Various isolation structures to reduce the impact of substrate coupling are examined by both TCAD (technology computer-aided design) simulations and by experimental measurements of dedicated test structures (Sect. 3.4).

1.2.3 Design of Fast-Switching Circuit Blocks

Chapter 4 proposes designs of fast-switching circuit blocks to control a buck converter with the required minimum on-time pulses as small as 3 ns for input voltages up to 50 V, and switching frequencies up to 30 MHz. A review of a voltage-mode control scheme of a buck converter gives an overview of the timing critical circuit blocks (Sect. 4.1). A PWM generator is proposed, which is able to generate on-time pulses smaller than 3 ns (Sect. 4.2). Level shifters are proposed for the control of a PMOS high-side switch with static gate driver supply, as well for the control of an NMOS high-side switch with a switching gate driver supply (Sect. 4.3). A gate driver design is described which is optimized to transferring minimum on-time pulses with a minimum propagation delay. The output stage allows a fast turn-on of the power switch with optimized power consumption (Sect. 4.4).

1.2.4 Efficiency and Loss Modeling of High-V_{in} Multi-MHz Converters

In Chap. 5, the main loss contributors with respect to high input voltages and fast switching are described. Models are derived, which are the basis for an implemented efficiency model. Existing efficiency models are reviewed with respect to their capability for high-V_{in} multi-MHz switching converters (Sect. 5.1). The loss contributors of a buck converter and their priorities for an accurate loss modeling are described (Sect. 5.2). The specific loss contributors for an asynchronous buck converter are analyzed. A four-phase model is developed, which allows a separation of loss causes and loss locations. The impact of parasitic capacitance and transition losses on the model accuracy is discussed and an improved model is proposed (Sect. 5.3). Additional loss contributors of a synchronous buck converter are shown. The loss impact of using a low-side switch (instead of a diode in an asynchronous converter) is discussed. The losses related to the required dead time are analyzed. Requirements for a dead time control, which allows to fully eliminate dead time related losses, are derived (Sect. 5.4). General potential loss optimizations are pointed out (Sect. 5.5). Buck converter architectures, including the use of a PMOS or NMOS high-side switch, are compared over a wide input voltage and load range with respect to efficiency (Sect. 5.6). A design indicator is proposed, which allows to benchmark an efficiency performance of converters independently of the operating point, which enables a comparison of state-of-the-art converters published at different operating points (Sect. 5.7).

1.2.5   Dead Time Control

In Chap. 6, an implementation of a dead time control is proposed to eliminate dead time related losses. Available dead time control concepts are compared (Sect. 6.1). The proposed dead time control is implemented as predictive mixed-signal dead time control with a resolution of 125 ps and a range of 32 ns. It comprises digitally controlled differential delay chain elements to achieve the required accuracy, and a high-speed sampling stage to predictively evaluate the dead time state for an adjustment in the following switching period (Sect. 6.2). An enhanced light-load operation mode is proposed, which enables the dead time control to achieve soft-switching at both the low-side and high-side switch turn-on at low output currents, resulting in an improved light-load efficiency (Sect. 6.3).

1.2.6   Resonant Converters

Chapter 7 discusses resonant converter concepts, which overcome the efficiency decrease of hard-switching converters by achieving soft-switching of the power switches over a wide load and input-voltage range. A study based on a quasi-resonant converter demonstrates that conventional resonant converters are not suitable for widely changing input voltages and output currents due to its largely varying switching frequency (Sect. 7.1). A parallel-resonant converter (PRC) is proposed, which overcomes these limitations. It consists of a resonant circuit attached to a conventional buck output stage. A mixed-signal soft-switching control, supported by a fully integrated adjustable 5 bit capacitor array, allows achieving soft-switching at both, the main switch and the resonant switch of the converter (Sect. 7.2).

1.2.7 Conclusion and Outlook

Chapter 8 summarizes and concludes the results of this book.

References

1. Alimadadi M, Sheikhaei S, Lemieux G, Mirabbasi S, Dunford WG, Palmer PR (2009) A fully integrated 660 MHz low-swing energy-recycling DC–DC converter. IEEE Trans Power Electr 24(6):1475–1485. https://doi.org/10.1109/TPEL.2009.2013624
2. Bathily M, Allard B, Hasbani F (2012) A 200-MHz integrated buck converter with resonant gate drivers for an RF power amplifier. IEEE Trans Power Electr 27(2):610–613. https://doi.org/10.1109/TPEL.2011.2119380
3. BBC (2016) Human 'drone taxi' to be tested in Nevada. https://www.bbc.com/news/technology-36478614
4. Bergveld HJ, Nowak K, Karadi R, Iochem S, Ferreira J, Ledain S, Pieraerts E, Pommier M (2009) A 65-nm-CMOS 100-MHz 87%-efficient DC-DC down converter based on dual-die system-in-package integration. In: 2009 IEEE energy conversion congress and exposition, pp 3698–3705. https://doi.org/10.1109/ECCE.2009.5316334
5. Hazucha P, Schrom G, Hahn J, Bloechel BA, Hack P, Dermer GE, Narendra S, Gardner D, Karnik T, De V, Borkar S (2005) A 233-MHz 80%–87% efficient four-phase DC-DC converter utilizing air-core inductors on package. IEEE J Solid-State Circuits 40(4):838–845. https://doi.org/10.1109/JSSC.2004.842837
6. Huang C, Mok PKT (2013) An 82.4% efficiency package-bondwire-based four-phase fully integrated buck converter with flying capacitor for area reduction. In: 2013 IEEE International solid-state circuits conference digest of technical papers, pp 362–363. https://doi.org/10.1109/ISSCC.2013.6487770
7. Ishida K, Takemura K, Baba K, Takamiya M, Sakurai T (2010) 3D stacked buck converter with $15\,\mu m$ thick spiral inductor on silicon interposer for fine-grain power-supply voltage control in SiP's. In: 2010 IEEE International 3D systems integration conference (3DIC), pp 1–4. https://doi.org/10.1109/3DIC.2010.5751437
8. Kudva SS, Harjani R (2010) Fully integrated on-chip DC-DC converter with a $450\times$ output range. In: IEEE custom integrated circuits conference 2010, pp 1–4. https://doi.org/10.1109/CICC.2010.5617588
9. Kumawat AK, Thakur AK (2017) A comprehensive study of automotive 48-volt technology. SSRG Int J Mech Eng 4(5):7–14
10. Li X, Jiang S (2017) Google 48V power architecture. Plenary talk at Applied Power Electronics Conference and Exposition 2017, Google
11. Li P, Bhatia D, Xue L, Bashirullah R (2011) A 90–240 MHz hysteretic controlled DC-DC buck converter with digital phase locked loop synchronization. IEEE J Solid-State Circuits 46(9):2108–2119. https://doi.org/10.1109/JSSC.2011.2139550
12. Lu D, Yu J, Hong Z, Mao J, Zhao H (2012) A 1500 mA, 10 MHz on-time controlled buck converter with ripple compensation and efficiency optimization. In: 2012 Twenty-seventh annual IEEE applied power electronics conference and exposition (APEC), pp 1232–1237. https://doi.org/10.1109/APEC.2012.6165976
13. Maity A, Patra A, Yamamura N, Knight J (2011) Design of a 20 MHz DC-DC buck converter with 84 percent efficiency for portable applications. In: 2011 24th International conference on VLSI design (VLSI Design), pp 316–321. https://doi.org/10.1109/VLSID.2011.37
14. Muoio D (2017) RANKED: The 18 companies most likely to get self-driving cars on the road first. https://www.businessinsider.de/the-companies-most-likely-to-get-driverless-cars-on-the-road-first-2017-4?r=US&IR=T

15. Neveu F, Allard B, Martin C (2016) A review of state-of-the-art and proposal for high frequency inductive step-down DC-DC converter in advanced CMOS. Analog Integr Circuits Signal Process 87(2):201–211. https://doi.org/10.1007/s10470-015-0683-z

16. Peng H, Pala V, Chow TP, Hella M (2010) A 150MHz, 84% efficiency, two phase interleaved DC-DC converter in AlGaAs/GaAs P-HEMT technology for integrated power amplifier modules. In: 2010 IEEE radio frequency integrated circuits symposium, pp 259–262. https://doi.org/10.1109/RFIC.2010.5477346

17. Rix J (2018) Ten electric cars we're excited about in 2018. https://www.topgear.com/car-news/electric/ten-electric-cars-were-excited-about-2018

18. Sanders SR, Alon E, Le HP, Seeman MD, John M, Ng VW (2013) The road to fully integrated DC-DC conversion via the switched-capacitor approach. IEEE Trans Power Electron 28(9):4146–4155. https://doi.org/10.1109/TPEL.2012.2235084

19. Schrom G, Hazucha P, Hahn J, Gardner DS, Bloechel BA, Dermer G, Narendra SG, Karnik T, De V (2004) A 480-MHz, multi-phase interleaved buck DC-DC converter with hysteretic control. In: 2004 IEEE 35th annual power electronics specialists conference (IEEE Cat. No.04CH37551), vol 6, pp 4702–4707. https://doi.org/10.1109/PESC.2004.1354830

20. Schrom G, Hazucha P, Paillet F, Rennie DJ, Moon ST, Gardner DS, Kamik T, Sun P, Nguyen TT, Hill MJ, Radhakrishnan K, Memioglu T (2007) A 100MHz eight-phase buck converter delivering 12A in 25 mm^2 using air-core inductors. In: APEC 07 – Twenty-second annual IEEE applied power electronics conference and exposition, pp 727–730. https://doi.org/10.1109/APEX.2007.357595

21. Sturcken N, Petracca M, Warren S, Carloni LP, Peterchev AV, Shepard KL (2011) An integrated four-phase buck converter delivering 1A/mm^2 with 700ps controller delay and network-on-chip load in 45-nm SOI. In: 2011 IEEE custom integrated circuits conference (CICC), pp 1–4. https://doi.org/10.1109/CICC.2011.6055336

22. Sturcken N, O'Sullivan E, Wang N, Herget P, Webb B, Romankiw L, Petracca M, Davies R, Fontana R, Decad G, Kymissis I, Peterchev A, Carloni L, Gallagher W, Shepard K (2012) A 2.5D integrated voltage regulator using coupled-magnetic-core inductors on silicon interposer delivering 10.8A/mm^2. In: 2012 IEEE International solid-state circuits conference, pp 400–402. https://doi.org/10.1109/ISSCC.2012.6177064

23. Villar G, Alarcon E (2008) Monolithic integration of a 3-level DCM-operated low-floating-capacitor buck converter for DC-DC step-down donversion in standard CMOS. In: 2008 IEEE power electronics specialists conference, pp 4229–4235. https://doi.org/10.1109/PESC.2008.4592620

24. Villar-Piqué G, Bergveld HJ, Alarcón E (2013) Survey and benchmark of fully integrated switching power converters: switched-capacitor versus inductive approach. IEEE Trans Power Electron 28(9):4156–4167. https://doi.org/10.1109/TPEL.2013.2242094

25. Wens M, Steyaert M (2009) An 800mW fully-integrated 130nm CMOS DC-DC step-down multi-phase converter, with on-chip spiral inductors and capacitors. In: 2009 IEEE energy conversion congress and exposition, pp 3706–3709. https://doi.org/10.1109/ECCE.2009.5316434

26. Wibben J, Harjani R (2007) A high efficiency DC-DC converter using 2nH on-chip inductors. In: 2007 IEEE symposium on VLSI circuits, pp 22–23. https://doi.org/10.1109/VLSIC.2007.4342750

27. Wu F (2017) 48V: An improved power delivery system for data centers. White paper, Wiwynn, Taiwan

Chapter 2
Motivation for High-V_{in} Converters and Fundamentals

Several applications were identified, which are required to generate component level supplies from input voltages in the range of 5 V up to 50 V and higher. In most of these applications the power is supplied either by a battery or by an intermediate DC voltage. A continuously increasing demand for higher power consumption results in a trend towards an increasing DC supply voltage. In Sect. 2.1, an overview of applications in focus of this book and their special requirements to the DC-DC converters is given. Section 2.2 describes and compares different ways to realize DC-DC voltage conversions. It covers different voltage converter types and describes their benefits for high-V_{in} multi-MHz converters. Section 2.3 summarizes the fundamentals of buck converters, and Sect. 2.4 reviews the state-of-the-art converters, available for high-V_{in} and multi-MHz switching.

2.1 Applications for High-V_{in} DC-DC Converters

2.1.1 Automotive and e-Mobility

A significant increase in power demand can be observed in automotive applications, mainly driven by the trend towards electric vehicles, e-mobility, and higher end-user functionality. The conventional 12 V automotive board net with a 12 V alternator currently reaches its limit due to increasing functions in safety, driver assistant systems and electrical support, or replacement of mechanical components in braking, steering, oil pump, turbocharger, and others. Often more than 100 electronic control units (ECU) can be found in modern cars. Most of the ECUs contain several DC-DC converters to supply the circuit components and systems-on-chip (SoC), which are micro-controllers, sensors, motor and valve drivers, network interfaces (LIN, CAN, FLEXRAY, etc.), radar transceivers, displays, radio receivers, and many others. Several DC-DC converters are required at the point-of-load in the Electronic control

© Springer Nature Switzerland AG 2020

J. Wittmann, *Integrated High-V_{in} Multi-MHz Converters*,

https://doi.org/10.1007/978-3-030-25257-1_2

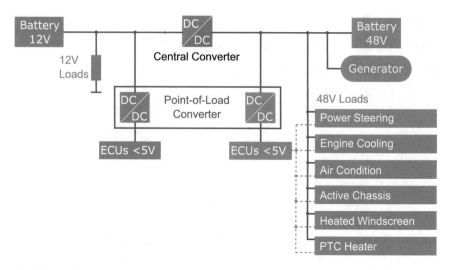

Fig. 2.1 Implementation of the 48 V board net, in addition to the conventional 12 V board net

units (ECUs) to generate various accurate voltage domains. The DC-DC converters often dominate the size and costs of the ECUs.

The power limitations of the 12 V board net are overcome by increasing the battery voltage. This allows to distribute the power at much lower currents and reduces the wiring costs. Additionally, the starter-generator can generate more electrical power at the same size if it operates at higher voltages. Therefore, a new standard for an automotive 48 V board net was introduced, and is becoming standard within the next years [7]. New functionalities, for example engine start/stop, energy recuperation, or faster engine start times, come together with a significant reduction of wiring and power consumption. Audi, for example, introduced the first 48 V-driven, fully electrical turbocharger in a serial production car [35], enabling the strongest diesel engine on the market. Currently, the 48 V board net is implemented in addition to the 12 V board net. A typical implementation is shown in Fig. 2.1. With a starter-generator typically operating at 48 V, a large central DC-DC converter transfers the energy between the 12 V and the 48 V battery. While the applications with lower power demand will remain at the 12 V battery to benefit from utilizing existing ECUs designed for 12 V supply, high-power applications are moved to the 48 V battery. However, it would be highly beneficial to supply all components in high-power ECUs directly from the 48 V board net, otherwise an additional wire would be required from the 12 V supply rail. Moreover, supplying one ECU from both board nets would require complicated and costly protection mechanism to prevent a short circuit between the two board nets in failure mode. In the future, the 12 V board net could be completely removed if it is possible to replace the 12 V DC-DC converters by 48 V converters without a significant impact on size, cost, and efficiency.

High-V_{in} DC-DC converters are also required in light-electric vehicles (LEV) and drones. As they are driven by a strong electric motor, a high battery voltage is required, with a voltage scaling with the size or power of the electrical vehicle. A most common battery voltage is 24 V or 36 V in e-bike, pedelecs, e-scooter, wheelchairs, electric sports vehicles (ESV), electric utility vehicles (EUV), and others, newer generations are also driven with a 48 V battery. Especially in LEVs, a conversion to an intermediate voltage of 12 V would not be accepted due to cost reasons. Consequently, high conversion ratios from up to 48 V down to below 5 V are required.

Larger hybrid or fully electric vehicles use batteries with voltages above 200 V, hybrid trucks even up to 700 V for the (hybrid) electric power trains. As high voltages above 60 V cannot be handled on the ECUs due to safety reasons, an intermediate supply rail at 12 V, 24 V (trucks), or even 48 V will be required in combination with a central high power DC-DC converter to transfer energy between the high voltage battery and the intermediate supply rail.

The requirements for the DC-DC converters are set by the applications. The priorities for the requirements are summarized in Table 2.1. The large number of DC-DC converters in a car results in a high cost pressure. Sometimes, the mechanical size of the application limits the size for the electronics and thus the converters. As the energy used for the electronics is only a fraction of the energy used for driving, converter efficiency has not the main priority. In a car, the majority of converters are still linear regulators with poor efficiency. Linear regulators often have to be replaced by more efficient switching converters, as the heat dissipation, caused by the converter losses, are determining the converter size and cost. Challenging for automotive converters is the wide input voltage range, which is caused by battery variations and over- and under-voltage conditions by load dumps. A high level of reliability is required, as in some cases the ECUs suffer from large temperature variations or vibrations during driving. A car is expected to operate for at least 15 years, but typically it is used only 5% of time. Functional safety has to be guaranteed, as a failing converter could lead to serious accidents.

Table 2.1 Priorities of the requirements for DC-DC converters for automotive and server applications (+: priority high, −: priority low)

Automotive		Requirements	Servers	
Low cost, small size	++	Cost/size	+	Converter costs vs. energy costs due to losses
Medium efficiency, cooling costs	+/−	Efficiency	++	Cooling, energy bill, worldwide energy consumption
12 V Battery: <6−40 V, 48 V Battery: 24–54 V	++	V_{in} range	−	5–10% of V_{in}
Operation: >15 years, 5% of time	+	Reliability	+	Operation: 2–4 years, 100% of time
Functional safety (ASIL)	++	Safety	−	None

In conclusion, further improvement towards full integration, also for conventional 12 V converters, is required to further reduce size, cost, and weight while maintaining an acceptable efficiency. New conversion techniques are required to cover very high conversion ratios for supply voltages up to 60 V and above, which allow a high level of integration at the same time [45].

2.1.2   Servers

The number of servers is significantly increasing. More than 11 million new servers were shipped worldwide every year in 2017, with the trend to increase by 5–10% per year [11]. Most of the servers are located in huge data centers [32], operated by internet service provider companies as Google, Facebook, Microsoft, Apple, Amazon, YouTube, Twitter, etc., and by companies offering data center services, for example, IBM, HP, Switch (SuperNAP Data Centers), Digital Reality, and many others. A strong increase in data traffic from 4.5 Zetabyte (=4.5 × 10^{15} MB) in 2015 to over ten Zetabyte in 2019 is expected by new trends like Cloud services, Internet of Things (IoT), and Big Data. Microsoft and IBM, for example, are about to construct warehouse scale data centers each larger than 500,000 m^2, three to five times larger than current world largest data centers. The total worldwide energy consumption by data centers was estimated to be in the range 250–350 TW h per year (2010), more than 1.5% of word wide energy consumption, equivalent to more than 50 coal power plants (500 MW) [20, 32]. Studies predict that the energy consumption by information and communications technology is about to double per decade [32], requiring an essential improvement of the power efficiency.

In a conventional data center facility, the energy is distributed by an AC distribution network as shown in Fig. 2.2a. An uninterruptible power supply (UPS) is supplied by the main grid or a backup generator with an AC voltage of typically 480 V. In the UPS, the voltage is converted to a DC voltage (e.g. 380 V) to charge large buffer batteries to supply the data center in case of a power loss at the grid. In conventional data centers, the voltage is transformed back to an AC voltage and is distributed in a power distribution unit (PDU) using transformer to generate a lower AC voltage, which supplies the particular server racks. To compare the power efficiency of data centers, the power usage effectiveness (PUE) was defined as a measure for the facility efficiency, while PUE = (facility power)/(server power), i.e., the power entering the data center facility in ratio to the power entering the server racks, containing the IT equipment. The average PUE of all data centers was around 1.65 in 2013 [2], i.e., in most of the data centers, less than half of the energy is used for the actual servers. Most losses are generated by the cooling system, while around one third of the losses are caused by the power distribution system. With better cooling systems and by changing the AC power distribution to a DC architecture, as shown in Fig. 2.2b, Google was able to reduce the PUE to 1.11 in 2017 [11], which is a loss reduction of about 70%. In data centers with optimized PUE, the focus of power loss optimization is directed towards the efficiency of the IT equipment on

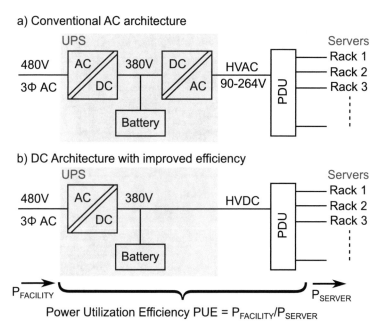

Fig. 2.2 Power distribution in a data center facility; (**a**) Conventional AC distribution architecture; (**b**) more efficient DC distribution architecture

rack level. Server utilization and power distribution within a rack is typically not included in the PUE metric.

A conventional power distribution on rack level is shown in Fig. 2.3a. The high-voltage supply (DC or AC) from the Power distribution unit (PDU) enters the rack and is distributed to up to ten server chassis with different sizes. The chassis offer the space for the server blades, containing the motherboards with CPUs, RAM, I/Os, or hard disks and other IT equipment. Conventionally, each chassis contains a power supply unit PSU, which generates a 12 V supply for several blades. In the blades, several voltage regulators (VR) generate the low-voltage supplies for the IT equipment at the point-of-load, with different voltages between 0.65 and 12 V. A server PUE (SPUE) metric was defined as the ratio of the power entering the server rack to the power consumed by the electronic components involved in computation (CPU, RAM, etc.). SPUE values in the range of 1.6–1.8 were common a view years ago, i.e., up to nearly half of the power in a server rack is lost, mostly by bad efficiency of the voltage converters [3]. Both conversion steps, in the PSUs and in the voltage regulators, used to have efficiencies below 80%. Implementing the HV-to-12 V converter in each chassis requires up to 10 power supply units (PSUs). This is costly and uses a large space in the server racks. A more efficient central 12 V converter for the entire rack shifts the problem towards the power distribution. Modern high-performance servers require up to 30 kW within one rack, which results in a supply currents of up to 2.5 kA to be distributed in a standard 9-inch

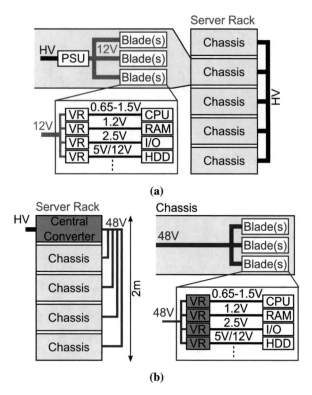

Fig. 2.3 Power distribution in a server rack; (**a**) Local 12 V conversion in each chassis; (**b**) more efficient central voltage conversion with a 48 V power distribution network on rack level

rack with a height of 2 m. Either costs for low-resistive copper wires or distribution losses become dominant.

To overcome this problem, Intel and Google proposed a rack level distribution system with a central HV-to-48 V converter, as shown in Fig. 2.3b [54]. A higher distribution voltage on rack level scales down the current. This is highly beneficial as the losses or the wiring costs are determined by the square of the distributed current. Compared to a 12 V distribution, a voltage increase to 48 V can reduce the cable costs to only 5% [19].

The scope of this book is on the improvement of voltage regulators suitable for the supply of the electrical components in the server blades, for both, a 12 and 48 V input. Besides the increasing input voltage of the voltage regulators, the output voltages are further reduced, as the component voltages for CPUs are decreasing towards 0.5 V or RAM towards 1 V to save computing energy. The conversion ratios of input-to-output voltage in the voltage regulators are becoming significantly higher. Section 2.2.1 shows that the converter efficiency also scales down with lower converter output voltages.

Further priorities for voltage regulators in servers are summarized and compared to automotive converters in Table 2.1. As future data centers will be mainly cost driven, more efficient converters will only be replaced if the lower energy bill with a higher efficiency pays off the higher converter cost and implementation effort. A high efficiency is important as it reduces cooling costs and the energy consumption of data centers at the same time. Due to the large amount of servers, the efficiency not only impacts the profit of the data center, but also the worldwide energy consumption. The output voltage of PSUs is regulated, and the computing load in servers is more defined as in automotive, consequently, the voltage converter's input voltage is rather constant. A typical lifetime of server equipment is up to 4 years. However, servers are often only allowed to be turned off a few minutes per year. This results in a total operating time, and thus reliability requirements, which is similar, or even higher, as in automotive, with the difference that no functional safety is required, as failing converters can be easily replaced.

2.1.3 Automation

With Internet of Things (IoT) and Industry 4.0, a high level of automation is the target in fabrication, logistic, or laboratory systems. Besides a high level of data exchange between robots and automated machines, the performances often relate to the number of objects to be processed or moved within a certain time, which mainly depends on the speed of robots or motors. Generally, the throughput is higher for smaller objects. To handle objects in, e.g., a production or assembly or line, electrical motors gain importance as they are more flexible and accurate compared to pneumatic systems. For heavy object, 3-phase AC motors, supplied from the grid, are required due to their high force. For lighter objects, low-voltage DC motors are often sufficient. Robotic systems with DC motors typically have a central AC-DC converter to generate a crude motor supply voltage. The motor controller sub-circuit, i.e., micro-controller, motor driver, sensors, etc., are supplied by down-converting the motor supply with local point-of-load DC-DC converters.

While the force of an electrical motor is obtained by the motor supply current (I_m) and the motor inductance (L_m), the acceleration mainly depends on the motor supply voltage (V_m), as the increase of the current over time is limited by L_m, with $V_m = L_m \cdot dI_m/dt$. Thus, the throughput of handled objects can be increased by a higher DC motor supply. This leads to a trend to increase the DC supply voltage up to the limit of the Safety Extra-Low Voltage (SELV) of 60 V [14] or 75 V [8] to avoid additional high-voltage safety protection at the motors. Consequently, highly integrated DC-DC converters for input voltages of up to 75 V converting to the low-voltage domains of the motor controllers, which are typically <5 V.

2.1.4 Mobile Devices

Mobile devices, as phones or tablets, are typically powered by battery voltages below 5 V. However, the capacity of the batteries often exceed 10 A h, while it is expected that the charging time decreases from generation to generation. A USB charger typically consists of a converter plugged to the mains, generating an intermediate voltage which is then supplying an internal charger IC in the mobile devices to power the battery. Long USB cables only can handle an increasing amount of current to be transferred. To achieve a higher charging power, the voltage of the USB charger port needs to be increased. Today, intermediate voltages up to 20 V are part of the USB Power Delivery standard [48], which is able to deliver up to 100 W of charging power. The USB charging voltage needs to be down-converted to about 4 V by a highly integrated charger IC to charge the battery of the mobile devices.

These applications demonstrated the trend and even the requirements towards higher supply voltages up to 50 V and higher, from which the power is delivered to the highly integrated voltage converters. The voltage converters have to cope with this trend to enable the increasing system supply voltages, rather than being their limitation.

2.2 Power Conversion to Low-Voltage Components

From the DC supply voltages offered by the applications, as described in Sect. 2.1, many kinds of low-voltage components have to be supplied with different input voltages levels in the range of 0.5–5 V. At least one highly integrated voltage regulator is required for each voltage rail. The voltage regulators are often integrated within the ICs on SoCs.

2.2.1 DC-DC Conversion Fundamentals

A DC-DC converter regulates a DC input voltage V_{in} to another DC output voltage output voltage (V_{out}). The output current (I_{out}) flows into the devices connected to the DC-DC converter's output. The connected devices or circuits represent a load R_{load} at V_{out}, consuming the converters output power P_{out}. An input current I_{in} flows from the input voltage V_{in} into the DC-DC converter's input. The input power P_{in} delivered to the input of the converter also has to contain the power losses generated in the converter. The input power results in $P_{in} = P_{out} + P_{loss}$. The power losses P_{loss} are dissipated as heat in the converter. High power losses require a high effort to dissipate the generated heat to keep the converter at an acceptable temperature.

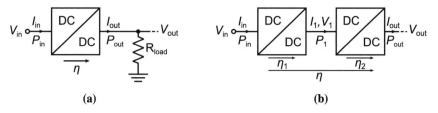

Fig. 2.4 DC-DC conversion from a high input voltage V_{in} to a lower output voltage V_{out}. (**a**) Basic, single-stage DC-DC converter. (**b**) Cascaded DC-DC converters

I_{out} determines the size of the power transistors or switches, which are required to regulate the output voltage at varying loads. Power transistors can be fully integrated up to currents of about 2–5 A. For output voltages up to 5 V, this results in an output power of up to 25 W. This book focuses on highly integrated converters with integrated power transistors (field-effect transistors, FET), implemented in a standard silicon technology.

The ratio between V_{in} and V_{out} is defined as the converter's voltage conversion ratio conversion ratio of a DC-DC converter (VCR), which is

$$VCR = \frac{V_{in}}{V_{out}}. \tag{2.1}$$

The power conversion efficiency η of a converter describes the ratio of P_{out} and P_{in}, as depicted in Fig. 2.4a. It is defined as

$$\eta = \frac{P_{out}}{P_{in}} = \frac{V_{out} \cdot I_{out}}{V_{in} \cdot I_{in}} = \frac{I_{out}}{VCR \cdot I_{in}}. \tag{2.2}$$

The efficiency η is the mostly used key parameter for converter comparison. Nevertheless, in some applications, the power efficiency is less important rather than the finally generated power losses, which determines the effort and the costs to dissipate the heat. From (2.2), it can be observed that the efficiency naturally scales down proportionally with the conversion ratio VCR. This is a fundamental problem for the trends to higher V_{in} and lower V_{out} in the applications offering high DC supply voltages.

For higher conversion ratios, e.g., $VCR > 5$, serial-connected (cascaded) DC-DC converters, as shown in Fig. 2.4b, are common [17]. Cascaded converters reduce the conversion ratio of each converter, and thus lead to an advantage in the implementation (see Sect. 2.3.6 and Chap. 4). While η_1 and η_2 (Fig. 2.4b) represent the conversion efficiency of each of the cascaded converters, the overall voltage conversion efficiency calculates to $\eta = \eta_1 \cdot \eta_2$. For example, a conversion from 12 to 3 V with a first converter with $V_{in} = 12\,V$ and $V_1 = 6\,V$, and a second converter with $V_1 = 6\,V$ and $V_{out} = 3\,V$, while each of the converters has an efficiency of $\eta_1 = \eta_2 = 90\%$, achieve an overall efficiency of $\eta = 81\%$. A single stage conversion, directly converting from V_{in} to V_{out} would only require

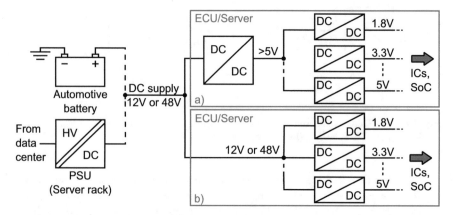

Fig. 2.5 Voltage regulation to low-voltage components, with (**a**) pre-regulation and (**b**) direct conversion

one converter with an efficiency of $\eta = 81\%$. As a conclusion, a single-stage converter might be the preferred choice, as it requires less efficiency, and only one converter might result in a smaller system size and cost compared to two cascaded converters. However, the implementation of converters with high conversion ratios has a limitation in switching frequency, and in size and cost. Thus the overall benefit depends on how small and how efficient converters with high-V_{in} and larger conversion ratios can be implemented. The limitations are discussed in Sect. 2.3.6.

Figure 2.5 shows two power distribution concepts with a pre-regulation (Fig. 2.5a) or a direct conversion (Fig. 2.5b). For higher DC supplies, e.g., 48 V, the pre-regulated concept is commonly used to reduce the conversion ratio of the converters. Anyway, a direct conversion would benefit from fewer converters (for more details see Sect. 2.3.2), and a lower efficiency can be accepted, compared to series connected converters in a pre-regulated system.

2.2.2   Types of DC-DC Converters

A DC-DC conversion can be performed by (1) a regulated resistor (linear regulators), (2) capacitively (switched-capacitor converters), or (3) inductively (inductive converters). A short overview of these conversion principles and an assessment of the capability for high-V_{in} is shown in the following.

2.2.2.1   Linear Regulators

An elementary type of DC-DC converters is a linear regulator. The principle is shown in Fig. 2.6a. With $I_{\text{out}} = I_{\text{in}}$, a lower V_{out} is achieved with a voltage drop

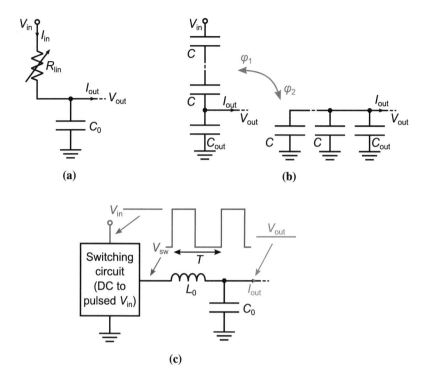

Fig. 2.6 Basic DC-DC conversion from V_{in} to a lower V_{out}. (**a**) Linear regulator principle. (**b**) Principle of a switched-capacitor (SC) converter. (**c**) Inductive converter principle

over a resistive element R_{lin}. The output voltage results in $V_{out} = R_{lin} \cdot I_{out}$, while R_{lin} adjusts the output voltage. R_{lin} is typically implemented as a regulated PMOS or NMOS transistor. The power dissipation over R_{lin}, which typically dominates the power dissipation of the entire linear regulator is $P_{loss} = (V_{in} - V_{out}) \cdot I_{out}$. The efficiency of a linear regulator is calculated as

$$\eta_{lin} = \frac{V_{out}}{V_{in}}. \tag{2.3}$$

As the power losses increase linearly with a rising V_{in} and a falling V_{out}, the power dissipation is very poor at high conversion ratios $VCR = V_{in}/V_{out}$. Thus, linear regulators are only appropriate for low conversion ratios if its low noise property or the low implementation cost is crucial, or very low efficiency can be accepted.

2.2.2.2 Switched-Capacitor Converters

Switched-capacitor (SC) converters perform a DC-DC conversion by switching several flying capacitors $C_{fly,n}$, once in series connection to V_{in} in a first phase φ_1, and once in parallel connection to V_{out} in phase φ_2, as shown in Fig. 2.6b. If the output current is zero, the voltage over each capacitor $C_{fly,n}$ equals the ideal output voltage $V_{out,ideal}$. This results in an ideal voltage conversion ratio N_i

$$N_i = \frac{V_{in}}{V_{out,ideal}}. \tag{2.4}$$

A load current I_{out} generates a voltage drop $V_{out,ideal} - V_{out} = I_{out} \cdot R_{sc}$ at the capacitors while they are connected to V_{out} in phase φ_2. R_{sc} represents the equivalent output resistance of the converter. R_{sc} and the voltage drop are determined by the size of the flying capacitors $C_{fly,n}$ and the switching frequency f_{sw}. The voltage drop from $V_{out,ideal}$ to V_{out} suffers from a low efficiency, which is identical to the linear regulator efficiency according to (2.3). Ideally, the efficiency of a switched-capacitor converter can be written as

$$\eta_{SC} = \frac{V_{out}}{V_{out,ideal}} = N_i \cdot \frac{V_{out}}{V_{in}} = \frac{N_i}{VCR}. \tag{2.5}$$

Theoretically, capacitive converters can achieve a very high efficiency at fixed conversion ratios close to the ideal conversion ratio N_i. For the applications covered in this book, a wide range of varying V_{in} (and thus VCR) has to be covered. At high VCR, large amount of separated capacitors $C_{fly,n}$ have to be connected in series, e.g., for $VCR > 10$ at least five separated capacitors are required [49]. Several switched-capacitor converters have been published, which are able to configure several different ideal conversion ratios N_i to cover larger V_{in} or V_{out} variations by switching the flying capacitors in various parallel-series combinations in phase φ_1 and φ_2 (see Fig. 2.6b). Figure 2.7 shows the theoretical efficiency (not covering switching and control losses) for a wide VCR range with three and eight different ideal conversion ratio (N_i). It can be observed that a higher number of N_i leads to a significant improvement of the efficiency minima, as the efficiency increases with a shorter distance to the next ideal conversion ratio N_i. Especially at high conversion ratios, a significant increase of the granularity of the ideal conversion ratios N_i is required to bring up the efficiency minima to an acceptable value. For increasing numbers of N_i, the switched-capacitor converter has to be able to connect the flying capacitors $C_{fly,n}$ into an increasing number of different combinations. The consequence is that several switches have to be connected to each side of $C_{fly,n}$ to realize the required capacitor combinations.

For a high efficient conversion over a large range of conversion ratios, tens of switches need to be implemented. Several of these switches are connected in series in the current path, limiting the maximum current of the converter due to a high switch resistance. Increasing the switches to reduce the resistance would result

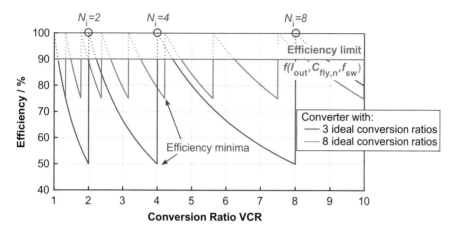

Fig. 2.7 Theoretical efficiency of a switched-capacitor converter at varying conversion ratios

in larger switching losses. At high V_{in}, most of the switches require high-voltage capability up to the input voltage, depending on the topology. Anyway, the gate control voltage is typically below 5 V. A large amount of independent flying voltage rails needs to be generated to control the switches. Usually, charge pumps or bootstrap circuits, level shifters, and an often area consuming high-side isolation for each switch add significant design effort and chip area. The area utilization for high-V_{in} converters is further reduced as high-voltage capacitors for $C_{fly,n}$ have to be used, which are typically low-density metal-metal capacitors.

At any of the converter's ideal conversion ratios N_i, the ideal output voltage $V_{out,ideal}$ cannot be reached as I_{out} leads to a voltage drop over the flying capacitors within each switching period, as they have to deliver charge to the output. This is equivalent to a voltage drop over the equivalent resistance, which is illustrated in Fig. 2.7. The switched-capacitor converter achieves a high peak efficiency at output voltages only very close to $V_{out,ideal}$ (or N_i). To increase the operating range of V_{out} towards $V_{out,ideal}$ and thus to increase the peak efficiency, the size of the flying capacitors or the switching frequency has to be increased. Larger $C_{fly,n}$ impacts chip area, while the frequency increase allows to cover the operating range towards ideal output voltage ($V_{out,ideal}$). However, higher switching frequency also limits efficiency, as it provokes large switching losses at the high-voltage switches. The limitation at the operating range of V_{out} and thus the peak efficiency is indicated in Fig. 2.7.

As a conclusion, the performance of switched-capacitor converters is expected to be superior at fixed conversion ratios with a low and stable input voltage, as well as small output currents. Wide-V_{in} switched-capacitor converters with $V_{in} > 10$ V have been published at an output power in the range of tens of milliwatts [30, 40].

2.2.2.3   Inductive Converters

The principle of an inductive converter is shown in Fig. 2.6c. A switching network, connected to V_{in}, generates a pulsed signal at the switching node V_{sw} with the amplitude of V_{in}. This signal is passed through a low-pass filter consisting of an inductor L_0 and a capacitor C_0 to filter the pulsed signal to a lower DC output voltage V_{out}.

By modulating the duty cycle of the pulsed signal at V_{sw} in a pulse-width modulated (PWM) fashion, the output voltage can be easily adjusted for a widely varying input voltage range, without changing the converter configuration.

Inductive converters achieve a significantly higher efficiency than linear regulators. Assuming the switches of the switching network and the output filter (L_0, C_0) as ideal, no losses occur at all. If any of the switches of the inductive converter is turned on, the switch carries the inductor current I_{L0}. Ideally, no voltage drop over the closed switch occurs, and the power losses of the switch are zero ($P_{loss} = 0\,V \cdot I_{L0}$). Also, no power losses occur if the switches are open, as the current is zero ($P_{loss} = 0\,A \cdot V_{sw}$). The reactive elements L_0 and C_0 store energy, which is transferred to V_{out} ideally without losses.

In a real implementation, several losses occur due to non-ideal elements. The switches have a remaining on-state resistance R_{on}. Inductors have a DC resistance R_{dcr}, core losses, losses by skin and proximity effect; capacitors have a series resistance R_{esr}, and parasitic capacitance of the switches and the inductor are present at the switching node, which have to be charged and discharged once in a switching period (details are described in Sect. 2.3.5 and Chap. 5). The losses can be scaled down by a lower switching frequency what requires larger filter components. The scalability between efficiency and size/cost allows a high level of on-chip integration.

The most widely used inductive converter type for the applications in focus of this book is the buck converter. In addition to the filter network, only two switches are required. The buck converter is one architecture, which is studied in this book, as it shows the highest potential for high integration with acceptable efficiency, even at high V_{in}.

The fundamentals and the relationship of the converter parameters, efficiency, and the size/cost are discussed in Sect. 2.3. The design and implementation of fast-switching buck converters at high V_{in} are shown in Chap. 3.

2.2.2.4   Soft-Switching and Resonant Converters

Soft-switching includes switching techniques, which are utilized to reduce switching losses [25]. Typically, they are applied in resonant or quasi-resonant converters [9, 34]. A theoretical explanation of different switching conditions, which are hard-switching, zero-voltage switching (ZVS), and soft-switching is shown in Fig. 2.8. An exemplary buck converter implementation with a high-side switch and a low-side Schottky diode is shown, including its parasitic capacitance C_{sw} at the switching

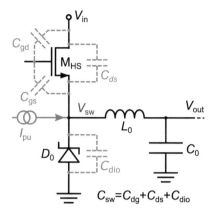

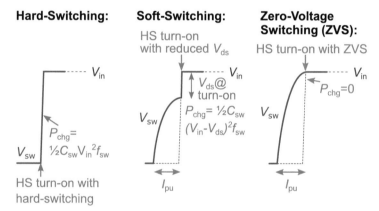

Fig. 2.8 Theoretical considerations and introduction to zero-voltage switching (ZVS) and soft-switching

node V_{sw}. In hard-switching condition, the high-side switch has to charge up C_{sw} from zero to the converters input voltage V_{in}, which causes significant charging losses P_{chg} as C_{sw} is charged resistively. The capacitive losses are covered in detail in Sect. 2.3.5 and Chap. 5.

The high-side switch is considered to turn on with soft-switching if the switching node V_{sw} is already pulled-up and C_{sw} are pre-charged to a higher voltage, shortly before the high-side switch turns on. The pull-up is caused by an additional circuit, like a resonant circuit or an inductor, which is able to provide a loss-less pull-up current I_{pu} for a very short time, and delivers sufficient charge to charge up C_{sw} as close as possible to V_{in}. This way, the remaining drain-source voltage V_{ds} at the high-side switch turn-on is significantly reduced compared to the hard-switching condition with $V_{ds} = V_{in}$. A reduction of the charging losses occurs. However, a loss-less generation of I_{pu} is the main challenge in soft-switching converters.

Zero-voltage switching (ZVS) at the high-side switch is the ideal case of soft-switching. The pull-up current I_{pu} is designed such that V_{sw} is charged up and reaches exactly V_{in}, when the high-side switch turns on with zero-voltage across V_{ds}. No additional charging of the switching node capacitance C_{sw} is required, and related losses are eliminated.

On top of the reduction of the charging losses, soft- or zero-voltage switching additionally reduced transition losses described in detail in Sect. 2.3.5 and Chap. 5. High transition losses occur, when the inductor current commutates to the high-side switch while V_{ds} is still large (e.g., $V_{ds} = V_{in}$ at hard-switching). The reduction of V_{ds} at turn-on by soft- or zero-voltage switching thus also reduces transition losses.

The benefit of zero-voltage switching is demonstrated in Fig. 2.9, which shows a break down of the converter losses of an asynchronous buck. The loss break down is obtained from a proposed efficiency model suitable for high-V_{in} multi-MHz switching (details see Chap. 5). It describes the loss causing elements of the converter, for example, the parasitic capacitances contributing to the switching node capacitance C_{sw}. While Fig. 2.9 (top) shows the losses in hard-switching condition, in Fig. 2.9 (bottom) the loss contributors are marked, which disappear if the converter is operated in ZVS (assuming an ideal I_{pu} to achieve ZVS condition). At an input voltage of $V_{in} = 12\,\text{V}$, the overall losses are theoretically reduced by 40%, while at higher input voltages, e.g., $V_{in} = 48\,\text{V}$, losses are even reduced by up to 80% with ZVS.

A synchronous buck converter, as described in Sect. 3.1, can additionally achieve soft- or zero-voltage switching at its low-side switch if an ideal pull-down current discharges the switching node from V_{in} towards ground, after the high-side switch is turned off and before the low-side switch is turned on. In a buck converter, this pull-down current is typically delivered intrinsically by a positive inductor current flowing out of the switching node. An adequate timing control of the turn-on and turn-off of the switches is required to achieve ZVS. In this book, this is addressed with a dead time control, described in Chap. 6. An enhanced dead time control proposed in Sect. 6.3 controls the low-side switch in a way to enforce negative inductor currents flowing into the switching node. At the low-side switch turn-off, the negative inductor current serves as pull-up current I_{pu} which charges up the switching node such that the high-side switch can be turned on with soft-switching as well.

Resonant converters are analyzed in Chap. 7. A parallel-resonant converter is proposed, which connects an auxiliary resonant circuit to the switching node of the buck converter, in order to generate the required pull-up current I_{pu} efficiently to enable soft-switching of the converter to increase its power efficiency.

2.2.2.5 Hybrid Converters

Hybrid converters are a combination of different converter architectures. Two main types of converters, which emerged in the past years, are multi-level buck converters

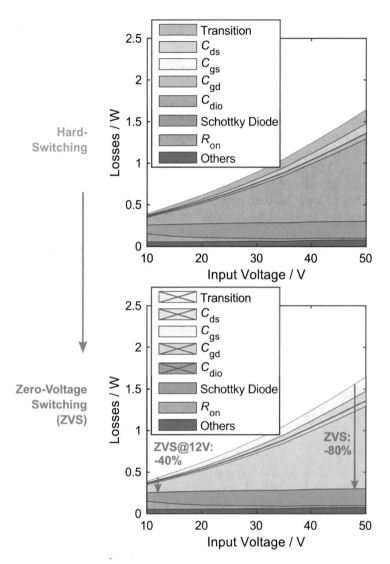

Fig. 2.9 Potential loss reduction in a synchronous buck converter with zero-voltage switching (ZVS) of the high-side switch, shown in a loss break down for an asynchronous buck converter with a switching frequency $f_{sw} = 10\,\text{MHz}$, a load current $I_{out} = 0.5\,\text{A}$, and an output voltage $V_{out} = 5\,\text{V}$

[10, 23, 24, 26–28, 51, 55, 56] and resonant switched-capacitor converters [18, 33, 36, 37, 41, 42].

Multi-level buck converters, especially three-level buck converters, combine a switched capacitor stage (see Sect. 2.2.2.2) into a buck converter (see Sect. 2.2.2.3). The switched-capacitor part performs a first conversion step with a fixed ratio. The remaining conversion ratio is realized by a buck converter. The benefit is a lower conversion ratio at the buck converter, allowing an operation at a reduced current ripple and a reduced voltage rating of the power switches. However, a three-level buck converter uses four switches, instead of two in a regular buck converter, which are in series. The converter operates with twice the on-state resistance, or twice the parasitic capacitance to be charged in each phase if the on-state resistance is kept equally to a regular buck converter. The disadvantages and advantages trade off and lead to a benefit of this converter only under specific condition. Multi-level converters are not covered in this book, but they are potentially suitable for high-V_{in} multi-MHz operation.

Resonant switched-capacitor converters add a small inductor to a switched-capacitor converter (see Sect. 2.2.2.2) to achieve a soft-switching in resonant mode (see Sect. 2.2.2.4). This converter type is also excluded from the scope of this book, as it is expected to have the limitations from both switched-capacitor and resonant converters, and thus might be only conditionally suitable for highly integrated high-V_{in} operation. A potential benefit is required to be investigated beyond the scope of this book.

2.3 Buck Converter Fundamentals

Section 2.3.1 gives an introduction to the most important parameters of the buck converter. In Sects. 2.3.2–2.3.4, the size and cost determining parameters are discussed. An overview of the loss causes is given in Sect. 2.3.5, and the challenges and limitations towards high V_{in} are shown in Sect. 2.3.6.

2.3.1 Buck Converter Parameters

The basic implementation of a buck converter is shown in Fig. 2.10. The switching circuit (see Fig. 2.6c) contains a high-side switch S_H and a low-side switch S_L which is implemented either as an integrated NMOS switch or a freewheeling diode (details see Sect. 3.1). The switches are controlled by the pulse-width modulated signals PWM and $\overline{PWM}$ to generate a pulsed input voltage V_{in} at the switching node V_{sw}. The filter L_0 and C_0 generates a lower DC output voltage V_{out}.

During the on-time t_{on} (see Fig. 2.10a), if PWM is high, S_H is conducting and V_{sw} is at V_{in}. During the off-time t_{off}, if PWM is low, S_L is conducting and V_{sw} is at ground. PWM has a switching period $T = t_{on} + t_{off}$ resulting in a switching

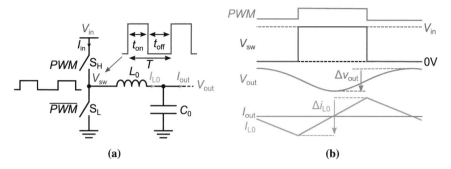

Fig. 2.10 Principle of a buck converter with (**a**) a buck converter schematic, and (**b**) the related signals

frequency $f_{sw} = 1/T$. The duty cycle D of PWM determines the voltage conversion ratio $VCR = V_{in}/V_{out}$, and thus the output voltage V_{out} of the converter, which is

$$D = \frac{V_{out}}{V_{in}} = \frac{1}{VCR} = \frac{t_{on}}{t_{on} + t_{off}} = \frac{t_{on}}{T} = t_{on} \cdot f_{sw}. \qquad (2.6)$$

Figure 2.10b shows the signals of the buck converter. During t_{on}, the voltage over L_0 is approximately $V_{L0} = V_{in} - V_{out}$. This leads to a linear increase of the inductor current I_{L0} within t_{on} by $\Delta i_{L0}(t_{on})$, which calculates to

$$\Delta i_{L0}(t_{on}) = t_{on} \cdot \frac{V_{in} - V_{out}}{L_0}. \qquad (2.7)$$

During t_{off}, the voltage over L_0 is V_{out}, as the primary inductor side is at ground. This leads to a linear decrease of the inductor current I_{L0} within t_{off} by $\Delta i_{L0}(t_{off})$, which calculates to

$$\Delta i_{L0}(t_{off}) = t_{off} \cdot \frac{V_{out}}{L_0}. \qquad (2.8)$$

If the actual output voltage V_{out} is lower than the target value resulting from the adjusted duty cycle D (2.6), the average inductor current increases from period to period, as $\Delta i_{L0}(t_{on}) > \Delta i_{L0}(t_{off})$. If the actual output voltage V_{out} is too large, the inductor current is decreasing accordingly. The converter is in steady state, when I_{L0} current increases and decreases by the same value within t_{on} and t_{off}, respectively ($\Delta i_{L0}(t_{on}) = \Delta i_{L0}(t_{off})$). The steady-state current ripple of the triangular like inductor current (as shown in Fig. 2.10b) can be expressed as

$$\Delta i_{L0} = \frac{(V_{in} - V_{out}) \cdot t_{on}}{L_0} = \frac{V_{out} \cdot t_{off}}{L_0} = \frac{V_{out}}{L_0 \cdot f_{sw}} \cdot \left(1 - \frac{V_{out}}{V_{in}}\right). \qquad (2.9)$$

The current ripple can be expressed as relative current ripple related to the output current, which is $\Delta i_{L0}/I_{out}$.

The output capacitor has to buffer the inductor current ripple, as the output current I_{out} is typically constant and do not follow the inductor current I_{L0}. The current I_{C0} flowing onto C_0 is $I_{C0} = I_{out} - I_{L0}$. I_{C0} generates a voltage ripple Δv_{out} on C_0, and thus at V_{out} (see Fig. 2.10b). The ripple Δv_{out} increases if I_{C0} is positive and decreases if I_{C0} is negative. Δv_{out} is calculated by integrating I_{C0} during the positive or the negative phase, respectively. Solving the integral during the time, when I_{L0} is positive, results in a positive voltage ripple

$$\Delta v_{out} = \frac{\Delta i_{L0}}{4 \cdot C_0 \cdot f_{sw}}. \tag{2.10}$$

Figure 2.11 sketches the influence of V_{out} and V_{in} on the current ripple Δi_{L0} and Δv_{out}. At constant V_{out} (Fig. 2.11a), both Δi_{L0} and Δv_{out} become larger with increasing V_{in}, however, both settle to an approximately constant value at conversion ratios VCR in the range of 5–10 (at V_{in} in the range of 25–50 V). At constant V_{in} and varying V_{out} (Fig. 2.11b), Δi_{L0} and Δv_{out} are small at low V_{out}, and increase towards higher V_{out} according to (2.9). Close to $VCR > 0.5$, the term $(1 - V_{out}/V_{in})$ in (2.9) becomes dominant and Δi_{L0}, as well as Δv_{out}, are decreasing again towards low VCR if V_{out} reaches V_{in}.

The consequence for high-V_{in} converters with high conversion ratio is a significantly higher inductor current and output voltage ripple, compared to low-V_{in} converters (low conversion ratio).

2.3.2 Scaling of the Output Filter

The size and the cost of a buck converter is mainly determined by the passive components L_0 and C_0, and the power switches. A relation of the particular converter parameters on the passive components is obtained by solving Eqs. (2.9) and (2.10) for the inductor value L_0 and C_0, which yields

$$L_0 = \frac{V_{out}}{\Delta i_{L0} \cdot f_{sw}} \cdot \left(1 - \frac{V_{out}}{V_{in}}\right) \tag{2.11}$$

and

$$C_0 = \frac{\Delta i_{L0}}{4 \cdot \Delta v_{out} \cdot f_{sw}}. \tag{2.12}$$

L_0 and C_0 are inversely proportional to Δi_{L0} and Δv_{out}. Thus, L_0 and C_0 scale identically with varying V_{out} or V_{in}, as it is depicted for Δi_{L0} and Δv_{out} in Fig. 2.11.

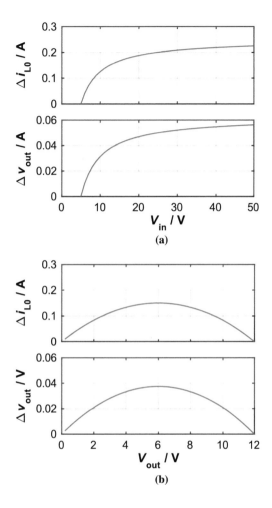

Fig. 2.11 Influence of V_{in} and V_{out} on the current and voltage ripple Δi_{L0} and Δv_{out} (with $L_0 = 2\,\mu H$, $C_0 = 100\,nF$, $f_{sw} = 10\,MHz$). (**a**) Variation of input voltage at $V_{out} = 5\,V$. (**b**) Variation of output voltage at $V_{in} = 12\,V$

Subsequently, the influence of the particular parameters on L_0 and C_0, and thus on the size of a buck converter, is described.

Output Voltage V_{out} linearly scales up L_0 if VCR remains unchanged. For example, a conversion from 50 to 10 V requires a 10× larger inductor than a conversion from 5 to 1 V. V_{out} has no influence on the value of C_0. Anyway, a higher V_{out} requires a capacitor with a higher voltage rating, which could influence the size and cost of C_0.

Input Voltage and Conversion Ratio If V_{in} increases at a fixed output voltage (increasing VCR), a larger value of L_0 is required, as long as VCR stays small. For larger conversion ratios ($VCR > 5$), the term $(1 - V_{out}/V_{in})$ becomes very small and the influence of V_{in} on L_0 disappears. As long as the converter is designed such that the current ripple remains unchanged, the size of C_0 is independent of V_{in}.

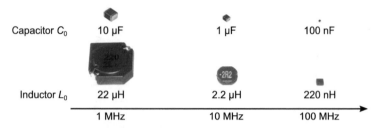

Fig. 2.12 Size reduction of C_0 and L_0 with increasing switching frequency

Switching Frequency An increasing f_{sw} linearly scales down both L_0 and C_0. Therefore, operating the converter at higher f_{sw} is the first choice if size and cost reduction is important. An example for size reduction of the passives with increasing switching frequency is demonstrated in Fig. 2.12. A detailed analysis of the inductor scaling with the switching frequency is done later in Sect. 2.3.3.

Inductor Current Ripple Δi_{L0} scales down L_0 but scales up C_0. Thus, the current ripple can be used to balance the values L_0 and C_0, i.e., the size of L_0 can be reduced at the expense of a larger C_0. To a certain extent, the inductor can be reduced by increasing the capacitor to balance the overall size and cost of the converter. However, Δi_{L0} can be increased only within the constraints of the converter efficiency. Smaller L_0 reduces the wiring and thus the inductor's DC resistance, however, larger Δi_{L0} increases both, DC losses (resistive losses in the switches and the inductor windings) and AC losses (core losses, skin and proximity effect). Thus, a properly chosen Δi_{L0} allows a further optimization of the converter efficiency and the size of the passives.

Output Voltage Ripple The output voltage ripple Δv_{out} linearly scales down with an increasing C_0 at a fixed Δi_{L0} and f_{sw}. For a better output voltage ripple, a larger C_0 is required. Alternatively, Δi_{L0} can be scaled down, resulting in a larger inductor value.

Output Current Both the values of L_0 and C_0 are independent of the output current I_{out}. Anyway, to keep losses low, a higher output current results in larger power switches with less on-state resistance, an inductor with lower R_{dcr} (thicker windings), and a core with a higher saturation current value.

2.3.3 Scaling of Real Inductors

Equation (2.11) shows that the value of an inductor for a given operating point V_{in} and V_{out} can be influenced by the switching frequency f_{sw} and the current ripple Δi_{L0}. In this section, the scaling of inductor losses, including AC losses (core losses, skin effect, proximity effect) and DC losses (wiring resistance), and the inductor volume towards higher switching frequencies is explored by a comparison of inductors available on the market. For this reason, inductors from a large inductor supplier are analyzed in terms of volume (and thus power density) and losses (with loss models provided by the supplier). The converter was assumed to operate at 48 V input, 5 V output, and 0.5 A load current.

First, the current ripple was chosen to be fixed at a typical value of $\Delta i_{L0} = 0.2\,A$ ($\Delta i_{L0}/I_{out} = 40\%$), and the inductors were chosen to operate at different switching frequencies. The required inductor value for each switching frequency was calculated. For each of the resulting inductor values, all suitable inductors, offered by the vendor, were analyzed by comparing inductor losses, inductor volume, and price. A summary of all analyzed inductors is shown in the Appendix (see Fig. 2.20). At each frequency, the inductor with the smallest available inductor volume is selected. Figure 2.13a (left) depicts the inductor losses, the inductor volume (marker size), and the price (color scale) of the selected inductors. It can be observed that a frequency increase allows scaling down the smallest available inductor volume, and at the same time it exhibits significantly lower inductor losses, as the DC resistance is reduced due to the reduction of the required inductor value. Only at frequencies above 30 MHz and at high current ripple, AC losses become dominant, and thus make air-core inductors the favorable choice. This is in accordance with a theoretical study of the inductor scaling in [39]. Moreover, the price of the inductors reduces towards smaller inductor volumes. At $f_{sw} = 1\,MHz$ the most expensive inductor is required, with a ferrite core of $L_0 = 22.4\,\mu H$, a volume of $17\,mm^3$, and a price of 0.78 US\$. Increasing the frequency, for example up to $f_{sw} = 20\,MHz$, the inductor value can be reduced to $L_0 = 1.1\,\mu H$ with a six times smaller volume of only $3\,mm^3$. This inductor is available at a price of only 0.30 US\$, even using a more expensive compound core material. Other operating conditions show a similar trend. Air-core inductors are not beneficial below 30 MHz, as increasing inductor windings without a core require thicker wires to keep the DC resistance low. This results in a significantly higher inductor volume and price, as the price generally scales with the amount of required wiring material (e.g., copper). Higher switching frequencies up to the range of 20–30 MHz are thus highly beneficial for smaller inductor volumes, lower price, and less heat dissipation.

Secondly, a similar inductor analysis is done, while the switching frequency is kept constant at $f_{sw} = 10\,MHz$ and the inductors values are selected to achieve different inductor current ripple. Again, the inductors with the smallest available inductor volume are depicted in Fig. 2.13a (right) for different current ripple. The inductor losses increase by more than three times if the current ripple is increased from below 100% towards 300% (at 0.5 A load) if the inductor volume remains equal. The inductor losses can only be decreased by significantly increasing the inductor volume, which is demonstrated in Fig. 2.13a (right) with two additional inductors of $0.3\,\mu H$. These inductors, which operate at a current ripple of 270%, are chosen such that they have inductor losses comparable to a $1.5\,\mu H$ inductor operating at 60% current ripple. Even at a five times smaller inductor value, the inductor volume is at least three times larger.

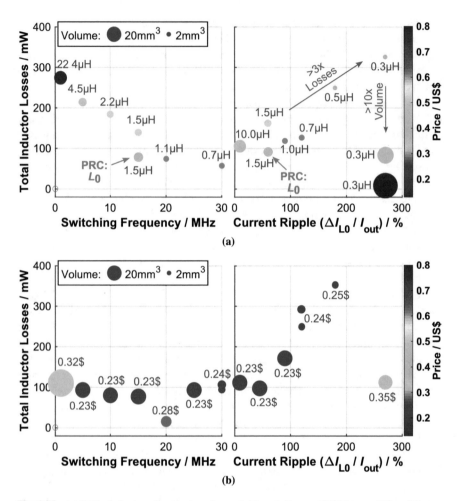

Fig. 2.13 Available inductors for varying f_{sw} and Δi_{L0} at $V_{in} = 48\,\text{V}, V_{out} = 5\,\text{V}$ and $I_{out} = 0.5\,\text{A}$. (**a**) Losses of selected inductors with minimum volume. (**b**) Losses of the inductors with minimum price

As an example, a selected inductor for a proposed converter of this book, which will be shown in Sect. 7.2, is added to Fig. 2.13a. The converter is suitable to operate at switching frequencies of 8–25 MHz. The output filter inductor (L_0) is chosen as an inductor with a ferrite core with a value of 1.5 µH, as this results in a current ripple range of approximately 80–380 mA (15–75% at $I_{out} = 0.5\,\text{A}$). According to the analysis of Fig. 2.13a, the current ripple allows to determine the best trade-off between inductor losses and inductor volume. By further increasing the current ripple with a slightly smaller inductor value (e.g., $L_0 = 1\,\mu\text{H}$), the inductor volume could be further reduced, with the drawback of higher inductor losses.

Figure 2.13b shows a similar analysis, while for each operating point the inductors with the lowest price are depicted. It can be observed that the minimum possible inductor price and the related volume is rather constant across the frequency range from about 5–25 MHz. For higher current ripples, the minimum possible price is also in the same range, however, the inductor volume scales down with the drawback of increasing inductor losses.

As a conclusion, an increasing switching frequency and a selection of the inductor to operate in a favorable current ripple range enables smaller inductor volumes and prices, as well as lower inductor losses, which partially compensate the higher switching losses towards higher switching frequencies.

2.3.4   Scaling of Real Capacitors

Equation (2.12) showed that the output capacitor C_0 for a give output voltage can be reduced with a higher switching frequency f_{sw} and a lower inductor current ripple Δi_{L0}. The volumes of the capacitors are limited to its standardized packages. To obtain the optimal capacitor concerning price or volume for a given f_{sw} and Δi_{L0}, all available ceramic SMD capacitors (which are typical for this kind of converters) with equal voltage ratings were extracted from a larger internet vendor. Figure 2.14 shows the price (y-axis) and the volume (size of markers) for different capacitor values (x-axis). Figure 2.14a shows selected capacitors, which have the lowest price at each of the depicted capacitor values, while in Fig. 2.14b, the capacitors are selected for the smallest available package for each value, and thus for the smallest volume. It can be observed that the price for capacitors in the range of 100 pF to 100 µF has an U-shaped behavior. The cheapest capacitors are available in the range

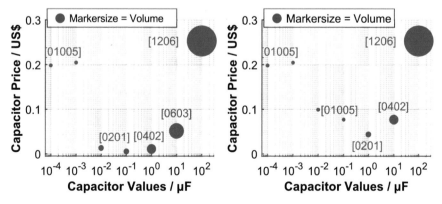

(a) Minimum price for different capacitors.   (b) Minimum volume for different capacitors.

Fig. 2.14 Available capacitors, suitable as a buck converter's output capacitor. (a) Minimum price for different capacitors. (b) Minimum volume for different capacitors

of 10 nF to 1 μF. Capacitors within large packages, as well as capacitors in very small packages are expensive. Figure 2.14 reveals that capacitors with a value of 100 nH and below are available in the smallest possible package. From this analysis it can be concluded that the optimal capacitor C_0 in terms of volume and costs is achieved, when the converter is designed such that C_0 is in the range of 10 nH and 1 μH. In certain operating points, a multi-MHz converter could be designed smaller and cheaper with a larger value of C_0, which even has a smaller output voltage ripple, or allows a higher Δi_{L0}, and thus a smaller value of L_0.

In conclusion, the costs and especially the size of the passives of a buck converter can be significantly reduced by increasing the switching frequency, and by choosing a suitable value for the inductor current ripple. While the converter even benefits from lower inductor losses at higher switching frequencies, the switching losses within the power stage significantly increase at higher f_{sw}, and impacts the overall converter efficiency.

2.3.5 Losses in a Buck Converter

The losses in a buck converter are reviewed by considering a real implementation of a synchronous buck converter, as depicted in Fig. 2.15. The ideal switches S_H and S_L from the principle buck converter schematic in Fig. 2.10a are assumed to be implemented as NMOS transistors. A more detailed loss consideration for different buck converter implementations is shown in Chap. 5. Figure 2.15 includes the parasitic capacitances of the NMOS switches. The parasitic capacitances appear between all terminals of the transistors. A gate driver circuit is required to rapidly

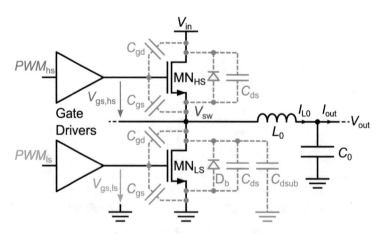

Fig. 2.15 Buck converter with parasitic capacitances, implemented with a synchronous output stage

charge the parasitic gate capacitances when the NMOS transistors are turned on or off.

The losses of the buck converter can be differentiated between static transition losses. Static losses occur during phases, in which the voltage and current conditions in the converter are nearly unchanged. Switching losses appear during the switching transitions, when the switches are turned on or off. An overview of the loss contributors is shown subsequently. Chapter 5 covers a more detailed analysis of these loss contributors and presents implementation of those as an efficiency model to simulate losses at varying operating points.

The following static losses occur in the converter [57]:

Switch conduction losses are caused by the inductor current flowing through the finite resistance of the power switches. The losses occur at the high-side switch during t_{on}, and at the low-side switch (or a freewheeling diode in an asynchronous buck converter, see Sect. 3.1) during t_{off}.

DC inductor losses are caused by the inductor current flowing through the DC resistance of the inductor windings.

The switching losses are categorized as follows:

Charging losses of the switching node capacitance occur at the converter's switching transitions. The parasitic capacitances C_{sw} at V_{sw} are either charged up to V_{in} when MN_{HS} turns on, or discharged to ground when MN_{LS} turns on. As the charging and discharging of C_{sw} is caused resistively by $R_{on,hs}$ or $R_{on,ls}$, losses occur in MN_{LS} and MN_{HS}.

Charging losses of gate capacitance are caused in the gate driver, when the power switch is turned on and turned off. Parasitic capacitances of the gate (C_{gd} including miller capacitance, and C_{gs}, see Fig. 2.15) of the power switches MN_{LS} and MN_{HS} are charged and discharged within the range of the gate driver supply voltage, which is typically in the range of 5 V in the technology used for this book. At $V_{gs,hs} = V_{gs,ls} = 5\,V$, the power switches are fully turned on. At $V_{gs,hs} = V_{gs,ls} = 0\,V$, the switches are turned off, i.e., $V_g(MN_{HS}) = V_{sw}$ and $V_g(MN_{LS}) = GND$. The gate charge at turn-on and turn-off of the power switches is dissipated by the driver transistors of the gate driver output stage to the gate driver supply or to the gate driver ground potential, respectively (details are shown in Sect. 4.4).

Transition losses occur, when a power transistor is turned on. The drain-source voltage V_{ds} over the switch drops from approximately V_{in} to very low voltages, as the parasitic capacitances C_{sw} at the switching node V_{sw} are charged or discharged with a nearly constant current. This current is limited by the transistor operating in saturation region during most of the transition time, especially at high V_{in}. This causes an approximately linear voltage transition at V_{ds}. As the inductor current I_{L0} fully started flowing through the transistor already at the beginning of the voltage transition, the power losses caused by the inductor current, which are $V_{ds} \cdot I_{L0}$, are significantly higher, as long as V_{ds} has not fallen to its final small value of the transistor's on-state.

AC inductor losses are mostly generated by the change in magnetization of a core material, which is not loss free. Thus, the stored energy in the inductor can be recovered entirely due to a hysteresis in the relation of the magnetic field intensity and the flux density. Hysteresis losses occur, which are proportional to the switching frequency. Other losses are caused by eddy currents, which are induced currents in the conducting core material. Both losses depend strongly on the design of the inductor core. Eddy currents are also responsible for increased losses in the inductor windings, which are known as skin effect and proximity effect.

Dead time related losses are related to the timing between the low-side switch turn-off and the high-side switch turn-on, and vice versa. The high-side and low-side switch in a synchronous buck converter are not allowed to be turned on at the same time, otherwise cross conduction occur, which lead to excessive losses. A dead time (also called non-overlap time) is required, i.e., the conducting switch is turned off before the open switch is turned on. Figure 2.16 demonstrates the dead time at the converter's turn-high transition. The gate drive signal of the low-side switch $V_{gs,ls}$ first turns off, while the gate driver signal of the high-side switch $V_{gs,hs}$ is turned on after a certain dead time. Assuming a positive inductor current, I_{L0} first drops over the on-state resistance of the low-side switch. During the dead time, in which none of the switches is conducting, I_{L0} commutates to the body diode D_b. The voltage drop over the low-side switch increases from $I_{L0} \cdot R_{on,ls}$ in switch conducting state to V_f during the dead time. V_f is the forward voltage drop over the body diode D_b while it is carrying I_{L0}. The voltage difference $(I_{L0} \cdot R_{on,ls}) - V_f$ generates additional losses caused by a dead time, which is chosen too long. When MN_{HS} turns on and V_{sw} is pulled high, the body diode D_b does not stop conducting immediately due to the reverse recovery effect. A large amount of charge is dissipated from V_{in} to ground as MN_{HS} and D_b are conducting at the same time, which lead to high reverse recovery losses.

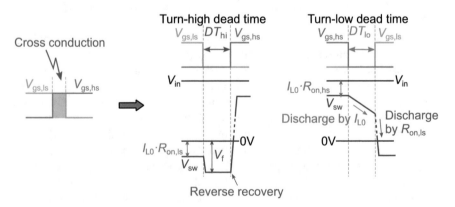

Fig. 2.16 Dead time and related losses

During the dead time at the converter's turn-low transition, the inductor current discharges the parasitic capacitances at the switching node after the high-side switch has turned off. This prevents a body diode conduction of the high-side switch. Thus, no diode losses occur in this case. As a non-resistive element (inductor) causes the discharge of V_{sw} and the charge of C_{sw} is recycled, as it is used as part of the output current I_{out}, ideally (assuming L_0 as ideal) no losses occur during the dead time. When MN_{LS} turns on after the dead time with a positive voltage at V_{sw}, the remaining charge at C_{sw} is dissipated to ground by $R_{on,ls}$ and losses occur in the low-side switch. The amount of charge dissipated to ground depends on the actual voltage at V_{sw} when MN_{LS} turns on. Ideally, dead time DT_{lo} is chosen long enough, such that I_{L0} is able to discharge C_{sw} to zero to eliminate the discharge losses, when the low-side switch is turned on. A dead time DT_{lo} chosen too short, significantly impacts the amount of losses occurring in the low-side switch. A detailed analysis of the dead time related losses is done in Sect. 5.4. To eliminate dead time related losses across varying operating points, a dead time control is proposed in Chap. 6.

Losses in gate driver and control circuits are mostly related to switching losses, as the dominant losses are caused by generating and propagating the frequency dependent PWM signal to the power switches. Especially the gate driver is accounting significantly to the losses due to the required driving strength for a fast turn-on of the power switches. The gate driver for the high-side switch requires an independent supply (high-side supply), as it is referenced to the source of the high-side transistor, and not to the overall system ground. As the PWM signal generation takes place on a low-side supply, which is typically 5 V or below, a level shifter is required to shift the PWM signal level to the high-side supply rails. The power consumption of each sub-circuits contributes to the overall losses of the converter. Low-power design techniques are required to limit the impact on the converter efficiency. Static losses of non-switching circuit blocks in the control loop and biasing circuits are typically not significant, as long as the converter is not operating at very low output power.

To optimize the converter losses, an accurate modeling of the different loss mechanisms is crucial. Chapter 5 shows the implementation of an efficiency model, which is specifically suitable to model the losses at high-V_{in} multi-MHz switching. It allows optimizing converter parameters such as the size of the power transistor, and enables a detailed analysis of the different loss contributors.

Figure 2.17 shows a typical loss distribution of a buck converter, obtained with the proposed efficiency model at $V_{out} = 5$ V and $I_{out} = 0.3$ A. In Fig. 2.17a, the converter losses are depicted over the switching frequency at $V_{in} = 48$ V. It can be observed that all switching losses are linearly increasing with the switching frequency and thus become dominant towards 10 MHz. Accordingly, the losses are depicted over the input voltage V_{in} for a switching frequency of $f_{sw} = 5$ MHz in Fig. 2.17b. In this simulation, it can be observed that the losses related to the switching node capacitance C_{sw} and also the dead time related losses even increase quadratically with the input voltage. The dead time is assumed to be fixed at a value of approximately 5 ns for the turn-high and the turn-low switching transition, which is a realistic value for integrated state-of-the-art converters. Due to charging

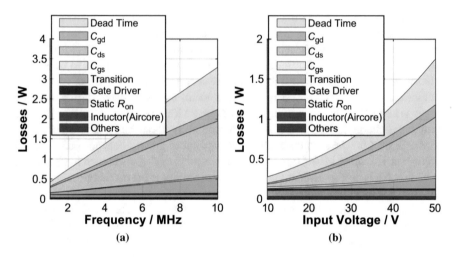

Fig. 2.17 Example for loss distribution in a buck converter with $I_{out} = 0.3\,A$, $V_{out} = 5\,V$ with (**a**) losses versus the switching frequency at $V_{in} = 48\,V$, and (**b**) losses versus the input voltage at $f_{sw} = 5\,MHz$

or discharging of the parasitic capacitances C_{sw} to or from V_{in} in each switching event, the dissipated energy in the power switch is

$$W_{sw} = \frac{1}{2} \cdot C_{sw} \cdot V_{in}^{2}. \tag{2.13}$$

The resulting power losses, which are $W_{sw} \cdot f_{sw}$, are thus proportional to V_{in}^{2}. The dead time related losses are depending linear on V_{in}. When the high-side switch turns on after the dead time DT_{hi}, the amount of charge Q_{rr} flowing through the low-side body diode during its reverse recovery time is delivered from V_{in} through the high-side switch. Thus, the resulting energy, which is lost in each high-side turn-on event is

$$W_{qrr} = Q_{rr} \cdot V_{in}. \tag{2.14}$$

The average power loss due to Q_{rr} calculates to $W_{qrr} \cdot f_{sw}$.

From Fig. 2.17, it can be concluded that a converter operation towards higher input voltages and an increasing switching frequency at the same time leads to significantly higher converter losses and lower power efficiency. An increase of the switching frequency, and thus related size of passives of a converter is limited by the converter losses. An optimization of the losses thus is related to enabling a smaller converter size.

The largest loss contributors at high V_{in} and high f_{sw}, which enable the highest potential for loss reduction, are losses caused by charging and discharging the switching node capacitance C_{sw}, transition losses, and losses related to the dead time.

In this book, an optimization of the converter is addressed in several aspects. (1) Different buck converter architectures, including technologies used for the power switches, are analyzed and compared with respect to power conversion efficiency and implementation effort. This allows discovering the most beneficial architecture for a given operating range of the converter. The trade-offs between different converter architectures are discussed in Sects. 3.1 and 5.6. (2) A dead time control to eliminate dead time related losses in an asynchronous buck converter is proposed in Chap. 6. (3) Achieving soft-switching condition at the power switches to reduce or even eliminate switching losses is addressed by resonant converters, which are discussed in Chap. 7. A parallel-resonant converter topology is proposed, which even allows soft-switching across a wide input voltage range.

2.3.6 Design Challenges for High-V_{in}

A fundamental limitation of the switching frequency of a buck converter at higher conversion ratios, and thus input voltages, is the minimum on-time of the PWM modulated switching signal V_{sw}. Figure 2.18a shows the timing of V_{sw} of a buck converter for different conversion ratios and switching frequencies. At 1 MHz switching, with a period of 1 μs, a voltage conversion ratio of $VCR = 2$ (e.g., $V_{in} = 4$ V and $V_{out} = 2$ V) requires a converter on-time $t_{on} = 500$ ns, as calculated from (2.6). Assuming an increase of the switching frequency to $f_{sw} = 30$ MHz, t_{on} significantly reduces to 15 ns. If the conversion ratio is further increased to $VCR = 10$, e.g., in a conversion from 50 V input to 5 V output, t_{on} reduces to a very small value of 3 ns at $f_{sw} = 30$ MHz. Already at lower input voltages, the on-time becomes critically small if the output is regulated directly to the point-of-load, e.g., in a conversion from $V_{in} = 12$ V to $V_{out} = 1.2$ V.

2.3.6.1 Circuit Design Requirements

To operate the switching node V_{sw} with an on-time as low as $t_{on} = 3$ ns, the according PWM signal has to be generated on the low-side domain, propagated through a level shifter to the high-side gate driver voltage domain (high-side) and finally control the high-side power switches via the gate driver. The gate driver has to be able to turn on the power switches within at least 1 ns. The low-side switch requires an inverted PWM signal, with a minimum off-time of 3 ns. All sub-circuits, which are responsible to generate or propagate the PWM signal are timing critical. They have to operate at the switching frequency, and they have to be able to propagate the minimum on/off-time pules without a significant disturbance of the pulse width. Circuit design for pulses in the low nanosecond range is critical for high-voltage converters, as a high-voltage technology has to be used for the sub-

Fig. 2.18 On-time limitation of the switching signal (V_{sw}) and its limitation by switching frequency, conversion ratio and input voltage. (**a**) Converter on-time t_{on} reduction with higher conversion ratios and higher switching frequency. (**b**) Technology limitation of switching transition sets minimum possible width of on-time pulse

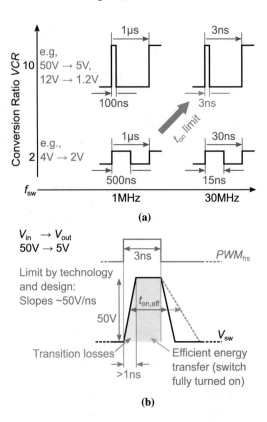

circuit design, which are typically optimized for high-voltage, while the regulation bandwidth is often limited to the sub-gigahertz range, allowing small pulse of a few nanoseconds only with special design techniques. The design of the circuit blocks complying to this requirement is proposed in Chap. 4.

2.3.6.2 Input-Voltage Dependent Limitation of the Switching Frequency

A second fundamental limitation of the switching frequency at high V_{in} is the limited rise time of the slope of the on-time pulse of the switching node V_{sw}, which is shown in Fig. 2.18b. Integrated power switches in silicon technologies allow a maximum voltage slope in the range of 50 V/ns. This value is independent of the power switch size, as larger switches would deliver a higher charging current, but the parasitic capacitances scale proportionally and thus lead to a constant voltage slope.

A realistic on-time pulse at $V_{in} = 50$ V, as it is depicted in Fig. 2.18b, would thus require at least 1 ns for the voltage transition from zero to 50 V. At a PWM on-time of 3 ns, a significant part of the converter energy is transferred during the switching transition, where it leads to high transition losses (a detailed transition loss analysis is covered in Sect. 5.3). The transition losses increase with shorter PWM on-time pulses, as a lower duty cycle is related to increase of the input voltage.

Concluding, an input voltage as high as 50 V suffers from a structural limitation of the switching frequency at around 30 MHz, depending on the technology. At lower V_{in} this limit scales up proportionally. The goal of this book is to demonstrate, how converters can be realized, which are operating up to this structural limitation.

2.4 State-of-the-Art Converters

A review of selection of relevant existing converters for input voltages up to 50 V with integrated power switches is shown in Fig. 2.19 [6]. The efficiency, the input voltage and the switching frequency of both, converters available on the market, and converters from scientific publications are compared. In Fig. 2.19a, the efficiency is shown versus the converter's input voltage, with the marker sizes representing the switching frequency. In Fig. 2.19b, the efficiency is plotted versus the switching frequency, with the marker sizes representing the input voltage.

Figure 2.19a shows that converters with integrated power switches are available for the entire input voltage range up to 50 V (and even higher). However, higher switching frequencies above 5 MHz can only be found in scientific publications and only at very small input voltages below 5 V [1, 4, 5, 12, 13, 15, 16, 21, 22, 29, 31, 38, 43, 44, 46, 47, 50, 52, 53]. Towards larger input voltages, converters operate at a decreasing switching frequency as well as a lower efficiency.

Figure 2.19b only contains converters with input voltages above 5 V. It can be seen that for higher V_{in}, switching frequencies are limited to below 5 MHz. As

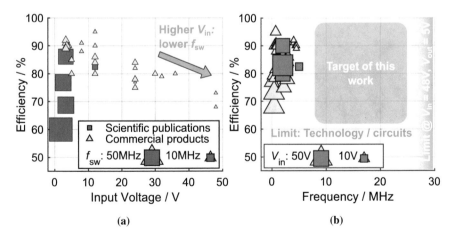

Fig. 2.19 Efficiency of published state-of-the-art converters and converters available on the market depicted over (**a**) the input voltage V_{in} and (**b**) the switching frequency f_{sw} for converters with $V_{\text{in}} > 5$ V

discussed in Sect. 2.3.6, the expected structural limitation by the technology is only expected to appear at around 30 MHz at, e.g., $V_{\text{in}} = 48$ V and a voltage conversion ratio $VCR = 10$. At lower input voltages or conversion ratios, this limitation is even shifted to higher switching frequencies. The frequency range between 5 MHz and the technology limit is not covered by any of the converters. This gap reveals a high potential for a significant reduction of the converter size and cost.

This book demonstrates how the frequency gap from available converters towards the structural technology limit can be closed or significantly tighten in order to achieve a significantly higher level of integration by faster switching. This book shows, how converters are pushed to the technology limit by (1) improving the limited speed of the circuit blocks to be able to realize and propagate on-time pulses in the range of 3 ns and (2) to address the decreasing converter efficiency by improving losses in conventional converters and finding new more efficient architectures.

Appendix

Figure 2.20 depicts a summary of all buck converter's output inductors analyzed in Sect. 2.3.3. For each operating point, a large variety of different inductors are available. The best compromise between volume, size, and inductor losses have to be found, depending on the system requirements.

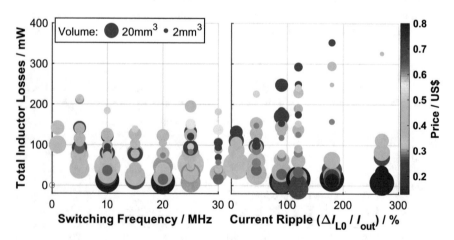

Fig. 2.20 Losses of selected inductors with a reasonable price or volume plotted versus f_{sw} at $\Delta i_{\text{L0}}/I_{\text{out}} = 40\%$ (left) and versus $\Delta i_{\text{L0}}/I_{\text{out}}$ at $f_{\text{sw}} = 40\%$ (right)

References

1. Alimadadi M, Sheikhaei S, Lemieux G, Mirabbasi S, Dunford WG, Palmer PR (2009) A fully integrated 660 MHz low-swing energy-recycling DC–DC converter. IEEE Trans Power Electron 24(6):1475–1485. https://doi.org/10.1109/TPEL.2009.2013624

2. Avgerinou M, Bertoldi P, Castellazzi L (2017) Trends in data centre energy consumption under the European code of conduct for data centre energy efficiency. MDPI AG, Basel

3. Barroso LA, Clidaras J, Hölzl U (2011) The datacenter as a computer, an introduction to the design of warehouse-scale machines, 2nd edn. Morgan and Claypool Publishers, San Rafael

4. Bathily M, Allard B, Hasbani F (2012) A 200-MHz integrated buck converter with resonant gate drivers for an RF power amplifier. IEEE Trans Power Electron 27(2):610–613. https://doi.org/10.1109/TPEL.2011.2119380

5. Bergveld HJ, Nowak K, Karadi R, Iochem S, Ferreira J, Ledain S, Pieraerts E, Pommier M (2009) A 65-nm-CMOS 100-MHz 87%-efficient DC-DC down converter based on dual-die system-in-package integration. In: 2009 IEEE energy conversion congress and exposition, pp 3698–3705. https://doi.org/10.1109/ECCE.2009.5316334

6. Commercial Devices (2015) Selection of available converters on the market: LT1076-5, LT3435, LT3481, LT3502, LT3505, LT3686, LT3689, LT3976, LT3990, LT8610, LTC3410, LTC3411A, LTC3561, LTC3564, LTC3565, LTC3568, LTC3601, LTC3604, LTC3612, LTC3614, LTC3616 (www.analog.com), TPS62615 (www.ti.com)

7. Els P (2015) 48 volt electrification: the next step to achieving 2020 emissions

8. European Union (2014) Directive 2014/35/EU of the European parliament and of the council of 26 February 2014. http://eur-lex.europa.eu/legal-content/EN/TXT/PDF/?uri=CELEX:32014L0035

9. Funk T, Wittmann J, Rosahl T, Wicht B (2015) A 20V, 8MHz resonant DCDC converter with predictive control for 1ns resolution soft-switching. In: 2015 IEEE international symposium on circuits and systems (ISCAS), pp 1742–1745. https://doi.org/10.1109/ISCAS.2015.7168990

10. Ganjavi A, Ghoreishy H, Ahmad AA (2018) A novel single-input dual-output three-level DC–DC converter. IEEE Trans Ind Electron 65(10):8101–8111. https://doi.org/10.1109/TIE.2018.2807384

11. Gartner (2018) Worldwide server shipments 2010–2017. https://www.statista.com/statistics/219596/worldwide-server-shipments-by-vendor/

12. Hazucha P, Schrom G, Hahn J, Bloechel BA, Hack P, Dermer GE, Narendra S, Gardner D, Karnik T, De V, Borkar S (2005) A 233-MHz 80%–87% efficient four-phase DC-DC converter utilizing air-core inductors on package. IEEE J Solid-State Circuits 40(4):838–845. https://doi.org/10.1109/JSSC.2004.842837

13. Huang C, Mok PKT (2013) An 82.4% efficiency package-bondwire-based four-phase fully integrated buck converter with flying capacitor for area reduction. In: 2013 IEEE international solid-state circuits conference digest of technical papers, pp 362–363. https://doi.org/10.1109/ISSCC.2013.6487770

14. IEC (12/2005) International standard IEC60950-1

15. Ishida K, Takemura K, Baba K, Takamiya M, Sakurai T (2010) 3D stacked buck converter with 15μm thick spiral inductor on silicon interposer for fine-grain power-supply voltage control in SiP's. In: 2010 IEEE International 3D Systems Integration Conference (3DIC), pp 1–4. https://doi.org/10.1109/3DIC.2010.5751437

16. Jiang Y, Fayed A (2016) A 1A, dual-inductor 4-output buck converter with 20MHz/100MHz dual-frequency switching and integrated output filters in 65 nm CMOS. IEEE J Solid-State Circuits 51(10):2485–2500. https://doi.org/10.1109/JSSC.2016.2588466

17. Ke X, Sankman J, Ma D (2016) A 5MHz, 24V-to-1.2V, AO^2T current mode buck converter with one-cycle transient response and sensorless current detection for medical meters. In: 2016 IEEE applied power electronics conference and exposition (APEC), pp 94–97. https://doi.org/10.1109/APEC.2016.7467857

18. Kiani MH, Stauth JT (2017) Optimization and comparison of hybrid-resonant switched capacitor DC-DC converter topologies. In: 2017 IEEE 18th workshop on control and modeling for power electronics (COMPEL), pp 1–8. https://doi.org/10.1109/COMPEL.2017.8013321

19. Kim D, He J, Figueroa DG (2016) 48V power delivery to Grantley reference board. Presented at the IEEE applied power electronics conference and exposition (APEC) 2016, Long Beach, USA

20. Koomey JG (2011) Growth in data center electricity use 2005 to 2010. Tech. rep., Stanford University, Stanford

21. Kudva SS, Harjani R (2010) Fully integrated on-chip DC-DC converter with a 450x output range. In: IEEE custom integrated circuits conference 2010, pp 1–4. https://doi.org/10.1109/CICC.2010.5617588

22. Li P, Bhatia D, Xue L, Bashirullah R (2011) A 90–240 MHz hysteretic controlled DC-DC buck converter with digital phase locked loop synchronization. IEEE J Solid-State Circuits 46(9):2108–2119. https://doi.org/10.1109/JSSC.2011.2139550

23. Ling R, Shu Z (2016) A piecewise sliding-mode controller for three level buck DC-DC converters. In: 2016 Chinese control and decision conference (CCDC), pp 643–648. https://doi.org/10.1109/CCDC.2016.7531064

24. Ling R, Shu Z, Hu Q, Song Y (2018) Second-order sliding-mode controlled three-level buck DC-DC converters. IEEE Trans Ind Electron 65(1):898–906. https://doi.org/10.1109/TIE.2017.2750610

25. Liu KH, Lee FCY (1990) Zero-voltage switching technique in DC/DC converters. IEEE Trans Power Electron 5(3):293–304. https://doi.org/10.1109/63.56520

26. Liu X, Mok PKT, Jiang J, Ki W (2016) Analysis and design considerations of integrated 3-level buck converters. IEEE Trans Circuits Syst I 63(5):671–682. https://doi.org/10.1109/TCSI.2016.2556098

27. Liu Y, Kumar A, Pervaiz S, Maksimovic D, Afridi KK (2017) A high-power-density low-profile DC-DC converter for cellphone battery charging applications. In: 2017 IEEE 18th workshop on control and modeling for power electronics (COMPEL), pp 1–6. https://doi.org/10.1109/COMPEL.2017.8013362

28. Liu X, Huang C, Mok PKT (2018) A high-frequency three-level buck converter with real-time calibration and wide output range for fast-DVS. IEEE J Solid-State Circuits 53(2):582–595. https://doi.org/10.1109/JSSC.2017.2755683

29. Lu D, Yu J, Hong Z, Mao J, Zhao H (2012) A 1500mA, 10MHz on-time controlled buck converter with ripple compensation and efficiency optimization. In: 2012 Twenty-seventh annual IEEE applied power electronics conference and exposition (APEC), pp 1232–1237. https://doi.org/10.1109/APEC.2012.6165976

30. Lutz D, Renz P, Wicht B (2016) A 10mW fully integrated 2-to-13V-input buck-boost SC converter with 81.5% peak efficiency. In: 2016 IEEE international solid-state circuits conference ISSCC, pp 224–225. https://doi.org/10.1109/ISSCC.2016.7417988

31. Maity A, Patra A, Yamamura N, Knight J (2011) Design of a 20 MHz DC-DC buck converter with 84 percent efficiency for portable applications. In: 2011 24th international conference on VLSI design (VLSI Design), pp 316–321. https://doi.org/10.1109/VLSID.2011.37

32. Mills MP (2013) The cloud begins with coal, big data, big networks, big infrastructure, and big power, an overview of the electricity used by the global digital ecosystem. Tech. rep., Techpundit, Houston

33. Moursy Y, Quelen A, Pillonnet G (2017) Challenges for fully-integrated resonant switched capacitor converters in CMOS technologies. In: 2017 24th IEEE international conference on electronics, circuits and systems (ICECS), pp 198–201. https://doi.org/10.1109/ICECS.2017.8292101

34. Mousavian H, Bakhshai A, Jain P (2016) An improved PDM control method for a high frequency quasi-resonant converter. In: 2016 IEEE energy conversion congress and exposition (ECCE), pp 1–8. https://doi.org/10.1109/ECCE.2016.7854846

35. Nikolic T (2016) Audi SQ7 debuts with world-first electric turbocharging. https://www.caradvice.com.au/422503/audi-sq7-debuts-with-world-first-electric-turbocharging/

36. Pasternak S, Schaef C, Stauth J (2016) Equivalent resistance approach to optimization, analysis and comparison of hybrid/resonant switched-capacitor converters. In: 2016 IEEE 17th workshop on control and modeling for power electronics (COMPEL), pp 1–8. https://doi.org/10.1109/COMPEL.2016.7556737

37. Pasternak SR, Kiani MH, Rentmeister JS, Stauth JT (2017) Modeling and performance limits of switched-capacitor DC-DC converters capable of resonant operation with a single inductor. IEEE J Emer Sel Topics Power Electron 5(4):1746–1760. https://doi.org/10.1109/JESTPE.2017.2730823

38. Peng H, Pala V, Chow TP, Hella M (2010) A 150MHz, 84% efficiency, two phase interleaved DC-DC converter in AlGaAs/GaAs P-HEMT technology for integrated power amplifier modules. In: 2010 IEEE radio frequency integrated circuits symposium, pp 259–262. https://doi.org/10.1109/RFIC.2010.5477346

39. Perreault DJ, Hu J, Rivas JM, Han Y, Leitermann O, Pilawa-Podgurski RCN, Sagneri A, Sullivan CR (2009) Opportunities and challenges in very high frequency power conversion. In: 2009 Twenty-fourth annual IEEE applied power electronics conference and exposition, pp 1–14. https://doi.org/10.1109/APEC.2009.4802625

40. Sarafianos A, Steyaert M (2015) Fully integrated wide input voltage range capacitive DC-DC converters: the folding Dickson converter. IEEE J Solid-State Circuits 50(7):1560–1570. https://doi.org/10.1109/JSSC.2015.2410800

41. Schaef C, Stauth JT (2018) A highly integrated series–parallel switched-capacitor converter with 12V input and quasi-resonant voltage-mode regulation. IEEE J Emer Sel Topics Power Electron 6(2):456–464. https://doi.org/10.1109/JESTPE.2017.2762083

42. Schaef C, Din E, Stauth JT (2017) A digitally controlled 94.8%-peak-efficiency hybrid switched-capacitor converter for bidirectional balancing and impedance-based diagnostics of lithium-ion battery arrays. In: 2017 IEEE international solid-state circuits conference (ISSCC), pp 180–181. https://doi.org/10.1109/ISSCC.2017.7870320

43. Schrom G, Hazucha P, Hahn J, Gardner DS, Bloechel BA, Dermer G, Narendra SG, Karnik T, De V (2004) A 480-MHz, multi-phase interleaved buck DC-DC converter with hysteretic control. In: 2004 IEEE 35th annual power electronics specialists conference (IEEE Cat. No.04CH37551), vol 6, pp 4702–4707. https://doi.org/10.1109/PESC.2004.1354830

44. Schrom G, Hazucha P, Paillet F, Rennie DJ, Moon ST, Gardner DS, Kamik T, Sun P, Nguyen TT, Hill MJ, Radhakrishnan K, Memioglu T (2007) A 100MHz eight-phase buck converter delivering 12A in 25mm^2 using air-core inductors. In: APEC 07 – Twenty-second annual IEEE applied power electronics conference and exposition, pp 727–730. https://doi.org/10.1109/APEX.2007.357595

45. Sridhar N (2013) Power electronics in automotive applications. Tech. rep., Texas Instruments, Dallas

46. Sturcken N, Petracca M, Warren S, Carloni LP, Peterchev AV, Shepard KL (2011) An integrated four-phase buck converter delivering 1A/mm^2 with 700ps controller delay and network-on-chip load in 45-nm SOI. In: 2011 IEEE custom integrated circuits conference (CICC), pp 1–4. https://doi.org/10.1109/CICC.2011.6055336

47. Sturcken N, O'Sullivan E, Wang N, Herget P, Webb B, Romankiw L, Petracca M, Davies R, Fontana R, Decad G, Kymissis I, Peterchev A, Carloni L, Gallagher W, Shepard K (2012) A 2.5D integrated voltage regulator using coupled-magnetic-core inductors on silicon interposer delivering 10.8A/mm^2. In: 2012 IEEE international solid-state circuits conference, pp 400–402. https://doi.org/10.1109/ISSCC.2012.6177064

48. USB Specification (2017) Universal Serial Bus Revision 3.2 Specification. USB 3.0 Promoter Group, rev. 1.0

49. Van Breussegem T, Steyaert M (2012) CMOS Integrated Capacitive DC-DC Converters. Springer Science & Business Media, Berlin

50. Villar G, Alarcon E (2008) Monolithic integration of a 3-level DCM-operated low-floating-capacitor buck converter for DC-DC step-down conversion in standard CMOS. In: 2008 IEEE power electronics specialists conference, pp 4229–4235. https://doi.org/10.1109/PESC.2008.4592620

51. Vukadinović N, Prodić A, Miwa BA, Arnold CB, Baker MW (2016) Extended wide-load range model for multi-level DC-DC converters and a practical dual-mode digital controller. In: 2016 IEEE applied power electronics conference and exposition (APEC), pp 1597–1602. https://doi.org/10.1109/APEC.2016.7468080

52. Wens M, Steyaert M (2009) An 800mW fully-integrated 130nm CMOS DC-DC step-down multi-phase converter, with on-chip spiral inductors and capacitors. In: 2009 IEEE energy conversion congress and exposition, pp 3706–3709. 10.1109/ECCE.2009.5316434

53. Wibben J, Harjani R (2007) A high efficiency DC-DC converter using 2nH on-chip inductors. In: 2007 IEEE symposium on VLSI circuits, pp 22–23. https://doi.org/10.1109/VLSIC.2007.4342750

54. Wu F (2017) 48V: an improved power delivery system for data centers. White paper, Wiwynn, Taiwan

55. Xue J, Lee H (2016) A 2MHz 12-to-100V 90%-efficiency self-balancing ZVS three-level DC-DC regulator with constant-frequency AOT V^2 control and 5ns ZVS turn-on delay. In: 2016 IEEE international solid-state circuits conference (ISSCC), pp 226–227. https://doi.org/10.1109/ISSCC.2016.7417989

56. Xue J, Lee H (2016) A 2MHz 12–100V 90% efficiency self-balancing ZVS reconfigurable three-level DC-DC regulator with constant-frequency adaptive-on-time V^2 control and nanosecond-scale ZVS turn-on delay. IEEE J Solid-State Circuits 51(12):2854–2866. https://doi.org/10.1109/JSSC.2016.2606581

57. Zhang W, Liu Y, Li Z, Zhang X (2009) The dynamic power loss analysis in buck converter. In: 2009 IEEE 6th international power electronics and motion control conference, pp 362–367. https://doi.org/10.1109/IPEMC.2009.5157413

Chapter 3
Fast-Switching High-V_{in} Buck Converters

This chapter compares different buck converter implementations with respect to high input voltages at high switching frequencies. It comprises different buck converter architectures, implementation aspects, an overview of power switch technologies, and the impact of fast switching on circuit blocks by switching noise.

3.1 Buck Converter Architectures

Two basic buck converter architectures are shown in Fig. 3.1, the synchronous buck converter using a freewheeling diode V_{dio} to offer a current path to GND when the high-side switch is off, and the synchronous buck converter using a power switch S_{L} instead of the freewheeling diode.

The asynchronous buck converter in Fig. 3.1a has the advantage that no control circuits are required for D_{L}. In comparison to the synchronous buck converter in Fig. 3.1b, the gate driver and the gate charge of S_{L} cause additional switching losses. Nevertheless, a power switch can be designed to have a low on-state resistance to reduce conduction losses of the inductor current, while the minimum voltage drop over a diode is always limited to its threshold voltage in the range of about 0.3–0.7 V, which leads to higher conduction losses. Using a low-side switch, a dead time needs to be introduced to avoid cross currents, as discussed in Sect. 2.3.5. Integrated Schottky diodes are often not available in many technologies, or they are not optimized for the requirements of a buck converter (e.g., high reverse recovery). Thus, freewheeling diodes are often placed as external component, while a low-side power switch allows to be fully integrated. An investigation, of which of the both architectures is superior, is done in Chap. 5 by considering all parameters impacting the losses.

A further property of an asynchronous buck converter is that the diode do not allow negative inductor currents, which can occur when the inductor current ripple

© Springer Nature Switzerland AG 2020
J. Wittmann, *Integrated High-V_{in} Multi-MHz Converters*,
https://doi.org/10.1007/978-3-030-25257-1_3

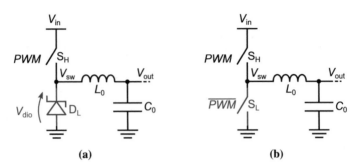

Fig. 3.1 Buck converter implementation: (**a**) Asynchronous buck converter with low-side (Schottky) diode; (**b**) Synchronous buck converter with low-side switch

exceeds the output current of the converter. If the inductor current is about to reverse, the diode stops conducting and the switching node becomes high impedance. This mode is called discontinuous conduction mode (DCM). Operating the converter in DCM has strong impact on the converter regulation and it is more challenging to achieve a stable output voltage regulation [4].

A synchronous converter allows to keep the switch turned on when the inductor current becomes negative, which leads to the continuous conduction mode (CCM). This is the preferred operating mode, as the regulation behavior does not change even for negative inductor currents. The option to operate the synchronous buck converter in DCM is possible by simply turning off the low-side switch at the point when the inductor current reverses.

A second fundamental decision impacting the architecture is whether a PMOS or an NMOS power transistor is used. A NMOS transistor typically has a better on-state resistance at lower parasitic capacitances, which significantly impacts the switching losses. For the low-side power switch in a synchronous converter, an NMOS transistor is always feasible, as its gate can be easily controlled between GND and the turn-on voltage which is 5 V in the technology used for the implemented converters of this book. This voltage is anyway required to supply the converter's low-voltage control circuits.

Using a PMOS as high-side transistor, as shown in Fig. 3.2a, has the advantage that the source of the PMOS transistor is connected to V_{in}, which is nearly constant within one switching period. Consequently, the gate-source control voltage of the PMOS switch (MP$_{hs}$) has to be referenced to V_{in}, and thus the gate driver ground $HSGND$, which is, e.g., 5 V below V_{in}, and the entire high-side supply rail is also constant over the whole switching period. Thus, the design criteria for the level shifter and the gate driver is, besides a low power consumption, that the PWM signal is transferred through the power stage (level shifter and gate driver) to the gate of MP$_{hs}$ with a minimum propagation delay $t_{d,ps}$ (Fig. 3.2a).

Using an NMOS transistor high-side switch, as shown in Fig. 3.2b, leads to a significant different behavior of the high-side supply rail, as the source of the NMOS transistor MN$_{HS}$ is connected to the switching node V_{sw}. The gate driver

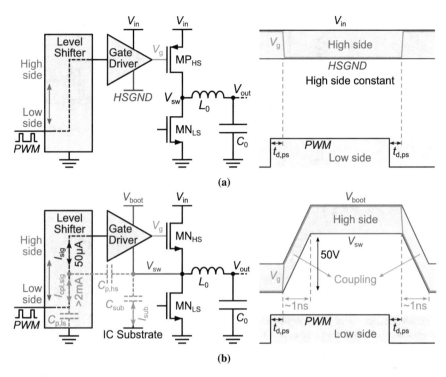

Fig. 3.2 High-side power switch in a buck converter implemented as (**a**) PMOS transistor and (**b**) NMOS transistor

ground and the output of the level shifter have to be referenced to V_{sw}, and a high-side gate driver supply V_{boot} has to be generated, which is, e.g., 5 V above V_{sw}. During switching events, the floating high-side supply rail follows the fast switching source of MN_{HS} with slopes up to 50 V within sometimes less than 1 ns. Parasitic capacitances at the high side have to be charged or discharged in each switching transition. High-side circuits in the level shifter contain parasitic capacitances from the high side to the signal path in the level shifter (represented by $C_{p,ls}$ in Fig. 3.2b), which cause coupling currents $I_{cpl,sig}$ into the level shifter of >2 mA at transitions as high as 50 V/ns. The coupling currents are overlaid with the signal currents $I_{cpl,sig}$ of the level shifter, resulting the signal currents to be disturbed and causing a false switching of the high-side power switch. Hence, the level shifter in a power stage controlling an NMOS switch additionally has to be very robust against these coupling currents.

Figure 3.3 shows different circuits to generate the high-side supplies. For a PMOS high-side switch, the most suitable circuits to generate the $HSGND$ rail is either a linear regulator (Fig. 3.3a) or a charge pump (Fig. 3.3b). Both can be integrated on chip. A linear regulator is superior in efficiency compared to a charge pump at lower V_{in}, as a charge pump causes switching losses and losses over the diodes. The pumping frequency of the charge pump could be decreased, which

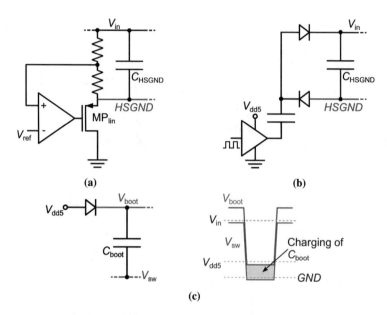

Fig. 3.3 Generation of the high-side supply voltages: (**a**) Linear regulator and (**b**) charge pump to generate $HSGND$ for controlling a high-side PMOS switch; (**c**) bootstrap circuit to generate a floating high-side supply for controlling a high-side NMOS switch

would require large capacitances, and prevent a full integration at least of the pumping capacitance. As the charge pump losses are not significantly depending on V_{in}, it is the preferred solution for higher V_{in}, where a linear regulator causes high losses due to a high voltage drop over the regulation transistor MP_{lin}.

The high-side supply, required to control an NMOS high-side switch, is typically generated by a bootstrap circuit, as shown in Fig. 3.3c [5]. While V_{sw} is at low level, C_{boot} is charged from the low-voltage domain V_{dd5}. During charging, the diode voltage drop V_{dio} is partially canceled, as the switching node is pulled below ground by the inductor current dropping over the on-state resistance of the low-side switch. The high-side voltage $V_{boot} - V_{sw}$ thus slightly depends on the converter output current and calculates to

$$V_{boot} - V_{sw} = V_{dd5} - V_{dio} + I_{L0} \cdot R_{on,ls}. \tag{3.1}$$

The buffer capacitors C_{HSGND} and C_{boot} are typically implemented off-chip. They have to be large to avoid a drop in the high-side supply, when the power switches are turned on or off, and a large amount of charge is required instantly to charge or discharge the gate of the power switches.

3.2 Power Switch Technologies

As the power switch is the most critical component in terms of converter size and efficiency, a comparison of suitable technologies for power switches is shown in Fig. 3.4. For input voltages up to 50–100 V, the power switch can be fully integrated as laterally diffused metal-oxide semiconductor (LDMOS) (Fig. 3.4a) in standard silicon (SI) CMOS technologies. This option allows a monolithic integration of the power switch on a single chip together with the control circuits of the voltage converter. This saves cost in packaging and has a benefit in very low parasitics especially in the gate control path, as the gate driver can be connected directly to the gate. However, the lateral implementation limits the current capability of LDMOS devices to a few amperes.

Power switches realized as a vertical structure are suitable for significant higher currents, as the available cross section for the current to flow is larger in vertical structures compared to lateral devices, in which the current mostly flows along the surface. A basic implementation is the double-diffused metal-oxide semiconductor (DMOS), which is shown in Fig. 3.4b. As this technology is not compatible to standard CMOS technologies, these switches have to be connected as separate

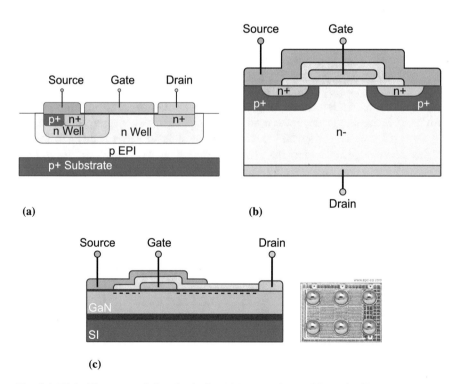

Fig. 3.4 High-side power switch technologies. (**a**) Integrated LDMOS FET in silicon. (**b**) External vertical switch in silicon. (**c**) External switch in gallium nitride (GaN)

devices to the controller IC of the voltage converter. This inherits significantly higher parasitics in the gate control loop due to the packages and bond wires.

A rather new technology for power switches is gallium nitride (GaN) [3, 7]. This technology allows a significantly higher power density and is especially suitable for fast switching [6, 10–15, 17, 20]. A cross section and a die photograph of a GaN power switch is shown in Fig. 3.4c. Attempts were made to make this technology compatible to standard CMOS, and to produce GaN devices on a SI waver. However, GaN on top of SI requires an additional technology [2]. Several standalone GaN devices are available on the market. Due to the fast switching capability, parasitics become even more crucial compared to vertical devices.

Some essential parameters, which are determining the switching behavior of the devices, are compared in Table 3.1. Lateral SI devices are typically integrated up to a maximum converter output current of about 5 A. External vertical SI metal-oxide field-effect transistors (MOSFET) are generally only available as larger devices suitable for small on-state resistance (e.g., <200 mV), which makes its usage only reasonable for converters with output currents above 5 A. External GaN switches are offered with a wide range of sizes ($R_{on} \ll 1\,\Omega$), which allows to cover large output currents of tens of amperes, as well as very small currents below 500 mA.

Crucial for the switching losses and switching behavior is the amount of charge Q_{oss} flowing into the drain (switching node) and the charge Q_g flowing into the gate of the devices in each switching event. Typically, Q_{oss} and Q_g determine the switching losses. Table 3.1 compares the figure-of-merits $Q_g \cdot R_{on}$, $Q_{oss} \cdot R_{on}$, as well as $Area \cdot R_{on}$, indicating the size of the switch at a certain R_{on}, for the three different switch technologies. Typical values of the figure-of-merits for external SI and GaN devices were previously published in [12]. In addition, these data were also extracted by simulation for an integrated 50 V LDMOS of a 180 nm high-voltage BiCMOS technology, which was used for the implemented converters of this book.

Table 3.1 shows that for a 50 V device, vertical SI devices require an up to 3 times higher gate charge and a more than 1.5 times higher on-state switching charge Q_{oss} as a GaN transistor with the same R_{on}. Also size of the device is about 1.5 times larger in vertical SI devices compared to GaN. The comparison of fully integrated

Table 3.1 Comparison of switching performance of different switch technologies used for a voltage converter with $V_{in} = 50$ V

	Lateral SI	Vertical SI	GaN
Switch	Full integration	External	External
$I_{out,max}$	<5 A	>5 A	>500 mA
$Q_g \cdot R_{on}$/nC mΩ	350	65	22
$Q_{oss} \cdot R_{on}$/nC mΩ	800	85	50
$Area \cdot R_{on}$/mm² mΩ	70	26	18
dV_{sw}/dt/V/ns	<50	<10[a]	>300[b]

[a]Typically limited by the parasitics of the package and internal wiring
[b]Slopes in published GaN converters are typically slowed down to <50 V/ns because of parasitics inductance

lateral SI devices resulted in a 5 times higher gate charge and nearly 10 times higher on-state switching charge compared to vertical SI devices, this is an up to 15 times higher Q_g and Q_{oss} compared to GaN.

Table 3.1 compares the maximum possible slope at the drain-source voltage (which equals the switching node voltage of the converter). Fully integrated lateral SI devices are able to achieve a switching slope of $dV_{sw}/dt < 50\,\mathrm{V/ns}$, measured on-chip. Data sheets of vertical SI MOSFETs typically state a maximum slope of less than $10\,\mathrm{V/ns}$, which is mainly limited by the package parasitics. Published measurements on GaN devices demonstrated switching slopes of several hundreds of V/ns [15]. This is an essential advantage, as the minimum possible on-time and thus the conversion ratio of a switching converter can be smaller, or the switching frequency can be further increased, as discussed in Sect. 2.3.6.

3.3 Converter Implementation Aspects

Fast switching transitions related to a fast power switch turn-on, which is required to overcome the minimum on-time limitations at high V_{in} and fast switching, as described in Sect. 2.3.6.2, result in a fast increase of the switching currents in the main power path, but also in the gate driver path. Fast increasing currents generate a large voltage drop over parasitic inductance. The critical current loops are depicted in Fig. 3.5 for both an asynchronous and synchronous buck converter. The switching current is provided by a large external buffer capacitance C_{in}, and delivered to the switch across the PCB and bond wire inductance L_{in}. In the asynchronous buck converter, shown in Fig. 3.5a, the return path of the current to C_{in} is across the switching node input and its parasitic inductance L_{sw}. In the synchronous converter in Fig. 3.5c, the return path is across the power ground path $L_{lp,gnd}$, as the low-side switch is fully integrated.

The gate charge currents delivered to the high-side power switch through the high-side gate driver are buffered by C_{boot}. The gate charge currents delivered to the low-side switch through the low-side gate driver are buffered by C_{vdd5} (synchronous converter only). A fast gate driver turn-on causes large voltage drops if these parasitic inductances are not negligibly small. An initial version of the PCB used for the asynchronous converter is shown in Fig. 3.6a, which was implemented according to the schematic in Fig. 3.5a. The IC was assembled in a conventional CSOIC28 IC package. The ringing due to the parasitic inductance is demonstrated exemplary for the high-side gate driver current loop by measuring the internal gate driver supply $V_{boot,i} - V_{sw,i}$ directly on the die, while the high-side power switch MN_{HS} is switching. The high-side supply rail is measured using low capacitive Picobrobes. The measured high-side supply voltage is shown in Fig. 3.6b. Especially at the turn-on event of the NMOS switch, a large ringing of up to $\pm 5\,\mathrm{V}$ is observed at the high-side supply rail. Negative peaks at $V_{boot,i} - V_{sw,i}$ nearly down to zero create a significant drop of the gate voltage of the power switch after MN_{HS} turned on, leading to an improper switching and higher losses. Positive peaks up to $10\,\mathrm{V}$

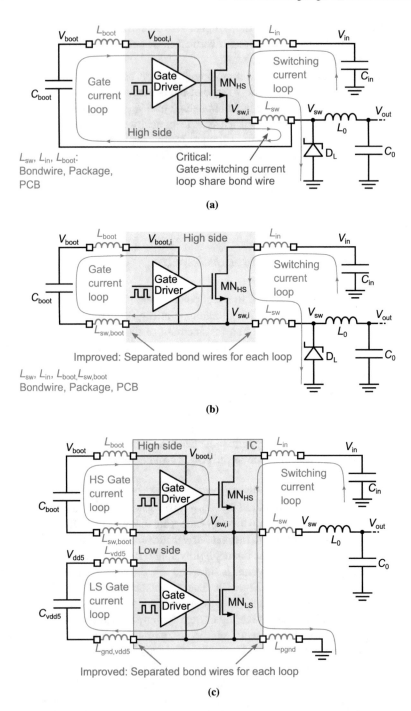

Fig. 3.5 PCB parasitics. (**a**) Un-optimized implementation of an asynchronous buck converter with a single bond at V_{sw} for both switching current loop and gate current loop. (**b**) Improved setup for an asynchronous buck converter, separating the gate current loop from the switching current loop with an additional bond wire at V_{sw}. (**c**) Improved PCB implementation for a synchronous buck converter

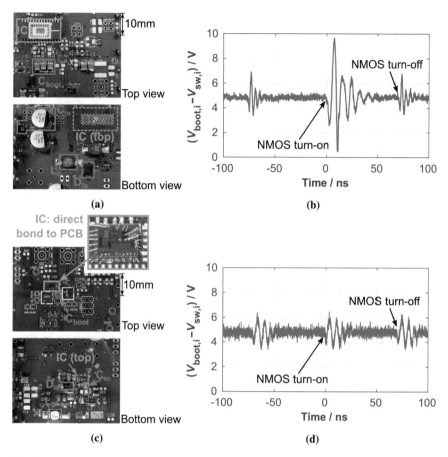

Fig. 3.6 (**a**) Initial PCB implementation resulting in (**b**) a large ringing at the high-side supply; (**c**) Improved PCB design with (**d**) significantly reduced ringing

exceeded the maximum ratings of the high-side transistors. Damages were observed due to electrical over-stress.

The ringing is mainly caused by (1) a large distance of the buffer capacitor C_{boot} of about 15 mm, also due to the bulky IC package. Increasing C_{boot} did not improve the ringing. (2) As depicted in Fig. 3.5a, the PCB design uses the same chip pin and bond wire L_{sw} for both, the connection of the switching node (switching current loop), and the bottom plate connection of C_{boot} (gate current loop). The voltage drop across L_{sw} is caused by the switching current loop while large load currents commutate fast from the freewheeling diode to the high-side switch. The gate current loop is also affected by this voltage drop if it shares the same chip pin. Conventionally, ringing is controlled by reducing the switching slope using a limited gate driver strength. However, this is not possible in fast-switching converters, as it would result in a limitation of the minimum PWM on-time, which impacts the maximum conversion ratio at fast switching frequencies.

The ringing is reduced with an improved board design according to the schematic in Fig. 3.5b, which is shown in Fig. 3.6c. In this design, the IC was directly bonded to the PCB, which allows to eliminate IC package related parasitics. The benefit is that all buffer capacitances can be placed very closely to the IC, and the wiring length of, e.g., C_{boot} to the IC, including bond wires, is reduced to approximately 2 mm. C_{in} is even placed on the bottom side of the PCB directly below the V_{in} chip pin and PCB bond wire pad.

A second improvement is the separation of the switching current loop from the gate driver loop by introducing a separate bond wire $L_{sw,boot}$ at the bottom plate of C_{boot}. This way, the voltage drop at the parasitic inductance in the switching current loops are decoupled from the gate current loop (see Fig. 3.5b), resulting in a major improvement of the ringing in the measurement of $V_{boot,i} - V_{sw,i}$, depicted in Fig. 3.6d. The improved PCB design including the direct bond technique reduces the ringing to approximately ± 1 V, which is a 80% reduction compared to the initial unoptimized PCB design (Fig. 3.6a). More advanced packages, e.g., waver level chip scale packages (WLCSP), enable a similar or even better performance.

The final implementation of the synchronous converter was improved accordingly. It requires an additional bond wire $L_{gnd,vdd5}$, to also separate the low-side gate current loop from the switching current loop, which is depicted in the schematic of Fig. 3.5c.

3.4 Substrate Coupling

The implementation of the integrated power stage of a buck converter with an NMOS FET as high-side power switch is shown in Fig. 3.7. As the source of an NMOS FET is connected to the switching node V_{sw}, the switching node forms a floating high-side reference ground to which the gate driver and the output of the level shifter are referred. The high-side circuits are implemented in an isolation well, which is doped complementary to the substrate.

The 180 nm high-voltage BiCMOS technology, used for the implemented converters of this book, contains a low-doped p^--epitaxial layer (EPI) on a high-doped p^+-substrate. Thus, the high-side isolation consists of a highly n^+-doped buried layer and deep medium doped n-well side-walls. The inner side of the high-side isolation is connected to V_{sw}, while the isolation structures (buried layer and deep n-wells) are connected to the high-side supply V_{boot}, which is typically 5 V above V_{sw}, supplied by an external buffer capacitor C_{boot} between V_{sw} and V_{boot}. V_{boot} defines the gate-source voltage of the NMOS FET at turn on. During the turn-on and turn-off transitions of the NMOS FET, V_{sw} is pulled high or low by a magnitude of several tens of volts within a view nanoseconds.

The isolation voltage V_{boot}, following V_{sw}, charges and discharges the parasitic capacitance C_{sub} (Fig. 3.7) of the depletion region between the outer n-doped

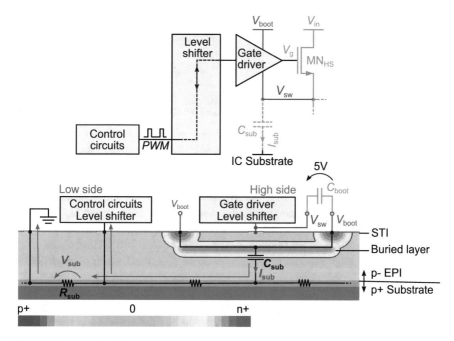

Fig. 3.7 Integrated power stage of a buck converter with NMOS high-side switch, implemented on a p^+/p^--substrate

isolation well and the p-doped EPI. A large amount of charge is injected into the substrate within a very short time. With rise times of V_{boot} up to $80\,V/ns$,[1] a parasitic capacitance of the isolation well to the substrate of a few picofarads causes coupling currents I_{sub} of several hundreds of milliampere. The current spreads across the substrate and is typically dissipated through substrate contacts and bond wires to the system ground GND. Substrate contacts are typically formed by very shallow moat contacts at the top of the highly resistive EPI. The high resistance along the path of the coupling currents R_{sub} causes a large voltage drop of up to several volts, strongly depending on the amount of substrate contacts and the quality of the connection of the contacts to GND. The voltage drop shifts the substrate potential of the wafer, and thus the reference ground of devices connected to the substrate. A first implementation of a fast-switching converter with just regular substrate contacts showed failures in the switching behavior caused by substrate coupling to the control circuits in the low-side voltage domain. Voltage and current references, but even digital signals, were observed to be disturbed.

In the following, dedicated isolation and low-resistive diverting structures suitable to minimize the voltage shift of the substrate potential are analyzed. The

---

[1]Observed in simulations without wiring parasitics. In measurements, maximum on-chip slopes in the range of 40–50 V/ns were observed.

structures are compared by TCAD simulations and measurements of implemented test structures with respect to effectiveness, area consumption, and implementation effort.

3.4.1   Comparison of Isolation Structures by Simulation

Several approaches to reduce substrate coupling have been published in the past. However, most of these publications analyze AC substrate coupling at very high frequencies, where the impedance of devices and the substrate is dominated by its capacitance [8, 9, 16, 18]. Digital circuits have much lower switching magnitudes and can be isolated to reduce coupling. Moreover, the noise is locally confined to very small sub-regions [1]. In fast switching power stages, the main energy is coupled to the substrate in the frequency range up to several 100 MHz at magnitudes of >50 V, and a large amount of charge has to be dissipated within one particular switching event.

To analyze different diverting structures, a 2.5D TCAD simulation with Synopsis Sentaurus was performed. Figure 3.8a shows the initially simulated doping structure, including the p^{+}-substrate, p^{-}-EPI, the high-side isolation well, and a reference current sink. To speed up simulations, only half of the structure was simulated, with a down-scaled well size of 70 μm × 70 μm. This is 50–100 times smaller than a typical high-side isolation in switching converters. However, realistic simulations matching the behavior of a converter implementation are expected, as the coupling effect is determined by the ratio between the well size and the area of the diverting structure. The sink in Fig. 3.8a is a reference current sink, which is connected to GND, consisting of a p-guard ring with two adjacent shallow p^{+}-moat contacts. It represents substrate contacts or devices using p-doped layers, distributed on the chip.

After a process simulation, a transient device simulation is performed, in which the isolation well is stimulated with a worst case voltage step of 40 V in 500 ps, which is the worst case condition seen in the transistor level simulation of the converter. Figure 3.8a–f show the doping (top) along with the current distribution of I_{sub} (bottom) at the time point of maximum coupling for various diverting structures.

The following structures have been investigated:

Basic doping profile without any specific diverting structure (Fig. 3.8a): The current is flowing directly down to the low-resistive p^{+}-substrate and flows back vertically across the EPI to the ground of the reference sink. Lateral coupling currents in the p^{-}-EPI are only significant up to a small distance of 50-100 μm away from the high-side isolation well (up to 4× the thickness of EPI) [19].

Moat contacts (Fig. 3.8b): Four stripes of standard substrate contacts (p^{+}-moat) surrounding the high-side isolation, each a width of 17.5 μm, are added to the basic doping structure. The moat contacts offer an additional current path for the coupling currents back to ground, which is still high resistive, as the moat contacts do not

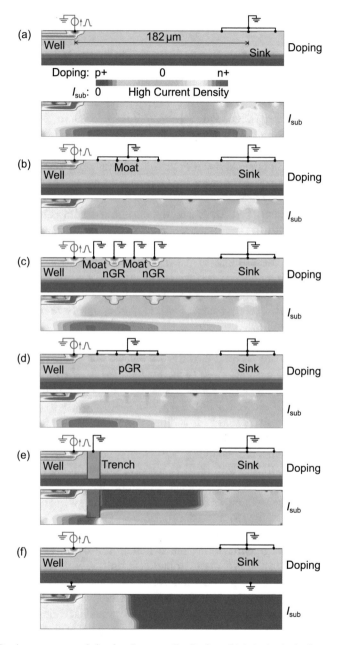

Fig. 3.8 Doping structure and simulated current distribution of (**a**) the basic doping structure, the structure with (**b**) moat contacts, (**c**) n-guard rings, (**d**) p-guard rings, (**e**) conducting trench and (**f**) back-side metalization BSM

reach deep into the EPI. The reference current sink still carries the major amount of current.

n-Guard rings (Fig. 3.8c): Due to the depletion region, standard n-guard rings (including adjacent substrate contacts) are only effective in one coupling direction, i.e. when the substrate is pulled below ground. As they are as deep as the isolation well, the idea of using n-guard rings is to block lateral coupling currents to adjacent low-voltage circuits. As lateral currents are minimal, n-guard rings are not effective.

p-Guard rings (Fig. 3.8d): The p-guard ring contains a p^+-buried layer under a p^+-doped deep well, which reduces the vertical resistance of the p^--EPI by ∼25%. As they do not extend down to the p^+-substrate, the high-resistive p^--EPI dominates the resistance in the current path. p-Guard rings achieve a suppression of the coupling current by 64%.

Conducting trench (Fig. 3.8e): The trench is filled with conducting material connecting down to the p^+-substrate, while the side-walls are covered with isolated oxide, to prevent lateral diffusion [18]. Nearly all the coupling currents are diverted by the conducting trench close to the high-side isolation well. Due to the low resistance, conducting trenches are much more efficient and consume less area than p-guard rings or moat contacts. Only few technologies offer conducting trenches, as they require additional process steps. For this reason, they have not been included in the experiments of Sect. 3.4.2.

Back-side metalization (BSM) (Fig. 3.8f): The coupling currents flow vertically to the substrate and are diverted directly at the back-side of the chip to the system ground. Only lateral currents close to the well at the top-side of the EPI influence low-side circuits, which can be further reduced by placing a minimal p-guard ring adjacent to the isolation well. BSM is not a standard process in most technology. As a post-processing step, BSM adds additional costs and packaging effort, and it is sometimes not favored due to known reliability issues.

To reduce the remaining coupling, low-voltage circuits can be protected with another isolation well. Only with very low resistive side-walls, the coupling from the outer well to the n-wells within the isolation can be diverted effectively. However, not all devices can be isolated, as the layers of the isolation well are used in some of the devices itself, e.g., bipolar transistors or high-voltage devices in the level shifter.

The critical coupling into low-voltage circuits mainly depends on the distribution of the substrate voltage shift during the switching transition. Figure 3.9 shows a comparison of the substrate voltage distribution during the high-side transition in the same order as in Fig. 3.8. Assuming the low-voltage circuits to be placed between the isolation well or diverting structure and the reference sink, the circuits would experience a substrate voltage drop of up to 3 V in the basic simulation structure (a), and up to 1 V with n-guard rings (b) or moat contacts (c). p-Guard rings (d) can reduce the voltage shift to less than 1 V. The conducting trench (e) shows a significant reduction to <500 mV, which could be further improved by increasing the width of the trench. Nearly no voltage shift in the substrate is seen for BSM (f).

Table 3.2 compares the different diverting structures with respect to reduction of the peak coupling current I_{sink} through the reference sink and the voltage drop across the substrate V_{sub} referred to the basic doping structure (Fig. 3.8a). Layout area and

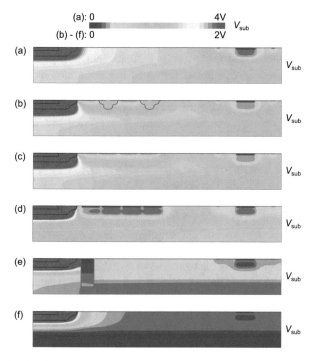

Fig. 3.9 Simulated voltage distribution of the investigated doping structures during the rising transition of V_{sw}. (**a**) Reference structure without additional isolation. (**b**) Standard substrate contacts (Moat). (**c**) n-Guard rings (nGR). (**d**) p-Guard rings (nGR). (**e**) Conducting trench. (**f**) Back-side metallization

Table 3.2 Comparison of diverting structures referred to unprotected substrate (TCAD results)

Structure	Improvement		Layout area	Additional costs
	I_{sink}	V_{sub}		
p^+-Moat contacts	54%	63%	−−	++
p-Guard rings	64%	75%	−	++
Conducting trench	92%	88%	+	+
Back-side metalization	>99%	>99%	++	−−

additional process costs are also covered in Table 3.2. Back-side metalization and conducting trenches are most effective structures to protect low-voltage circuits. If these technology options cannot be applied, p-guard rings are the best choice with the drawback that the area has to be scaled up significantly to achieve a sufficient voltage reduction. Compared to moat contacts, the area consumption of p-guard rings is reduced by one third and even further in technologies with higher doped and deeper p-layer.

3.4.2   Verification of Isolation Structures by Measurements

In order to verify the diverting capability of moat contacts, p-guard rings and BSM, two test chips were fabricated. The layout of the test chips, shown in Fig. 3.10a, contains an isolation well with a size according to a realistic application isolation well of 560 µm × 560 µm. The well is surrounded by two rings of moat contacts in one test chip and by two p-guard rings in a second test chip. A back-side metalization is realized with an aluminum coating in a post-processing step. To avoid a difficult current measurement on chip, a p-guard ring sink (Fig. 3.10a) was implemented. The sink is not connected to GND but is used as a low-resistive substrate contact to measure the voltage shift, caused by coupling currents, with a low-capacitive Picoprobe.

The measurement setup is shown in Fig. 3.10b. To obtain realistic measurement conditions, the isolation well of the test chips is bonded directly to the switching node of a separate buck converter chip. Parasitic inductances are reduced to a minimum by mounting both chips as close as possible to each other. The switching node voltage V_{sw}, applied to the test chips over a bond wire is depicted in Fig. 3.11a. It shows a maximum slope of up to 40 V/ns along with a significant ringing due to the bond wires. A ground connection with low impedance is realized by directly bonding the diverting structures to the PCB ground, or by connecting the back-side metalization to the ground plane using a conductive epoxy glue.

Figure 3.11b shows V_{sink} at 40 V/ns switching. A maximum substrate voltage shift of 500 mV is measured with MOAT1, 300 mV with p-GR1, and <80 mV with BSM, all connected to GND. The remaining voltage drop is caused by the parasitic inductance along the current path to the buffer capacitance C_{in} at V_{in} to GND, which is the source of the coupling currents. This remaining voltage drop is minimized by placing C_{in} very close to the chip to minimize bond wires and the connection path to the BSM. This confirms the coupling as predicted in

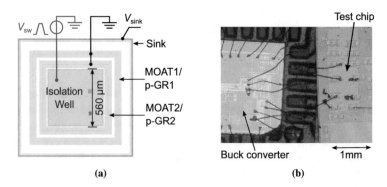

(a)   **(b)**

Fig. 3.10 Substrate coupling measurements. (**a**) Layout of the implemented test chips and (**b**) photograph of the measurement setup including the buck converter bonded to the test chip

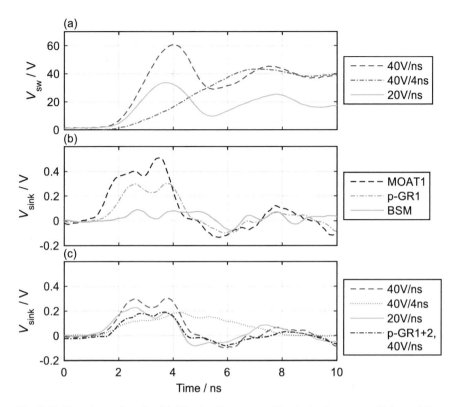

Fig. 3.11 Experimental results: (**a**) Stimulus V_{sw} generated by the buck converter; Voltage shift at V_{sink} for (**b**) 40 V/ns V_{sw} slope and for (**c**) varying magnitude and slope of V_{sw} and different guard ring area

Sect. 3.4.1 (Table 3.2). p-GR1 brings a reduction of around 30% versus MOAT1, without spending additional chip area.

The influence of the switching voltage magnitude and slope on V_{sink}, as well as of the area of the diverting structure is shown in Fig. 3.11c. The test chip with p-guard rings was measured at an identical slope of 40 V/ns, with 40 V and 20 V magnitude. As the isolation well capacitance behaves highly non-linear and increases towards lower well voltages, the major amount of charge is coupled at low V_{sw}. This explains that the coupling at half the magnitude is only reduced to 75%. A significant coupling already occurs at very low supply voltages. An additional measurement at 10 V/ns and 40 V supply shows how the disturbance reduces compared to 40 V/ns. Decreasing the slope by a factor of four decreases the voltage peak only to ∼65%, which is explained by the parasitic inductances of the connection of the switching node and ground and the non-linear well capacitance. It can be concluded that slopes of 1 V/ns are already critical. Another measurement uses both p-guard rings p-GR1 and p-GR2, which increases the effective p-GR area by 1.76 compared to p-GR1 only. The measured reduction follows almost exactly the p-GR area scaling of $1/1.76 = 57\%$.

As seen in Fig. 3.4a, also the drain of the an LDMOS transistor is connected to an n-well, which isolates the transistor. In a synchronous buck converter (see Fig. 3.1b), the drain of the low-side power switch connects to V_{sw}. In this case, the n-well connected to the drain also couples current into the substrate during high-side transitions. The coupling currents are in the same range as the coupling currents caused by the floating high-side isolation.

It has been shown that isolation and diverting structures are suitable and even mandatory to reduce coupling in fully integrated switched-mode power supplies at switching frequencies > 10 MHz with slopes > 1 V/ns and at various supply voltage levels, especially at 40 V and above.

References

1. Afzali-Kusha A, Nagata M, Verghese N, Allstot D (2006) Substrate noise coupling in SoC design: modeling, avoidance, and validation. Proc IEEE 94(12):2109–2138. https://doi.org/10.1109/JPROC.2006.886029
2. Boles T (2017) GaN-on-silicon present challenges and future opportunities. In: 2017 12th European microwave integrated circuits conference (EuMIC), pp 21–24. https://doi.org/10.23919/EuMIC.2017.8230650
3. Colino SL, Beach RA (2011) Fundamentals of gallium nitride power transistors. Application note: An002, Efficient Power Conversion Corporation, El Segundo
4. Erickson RW, Maksimovic D (2007) Fundamentals of power electronics. Springer, Berlin
5. Finco S, Tavares P, Fiore De Mattos A, Castro Simas M (2002) Power integrated circuit drives based on HV NMOS. In: 2002 IEEE 33rd annual IEEE power electronics specialists conference, vol 4, pp 1737–1740. https://doi.org/10.1109/PSEC.2002.1023061
6. Ganjavi A, Ghoreishy H, Ahmad AA (2018) A novel single-input dual-output three-level DC–DC converter. IEEE Trans Ind Electron 65(10):8101–8111. https://doi.org/10.1109/TIE.2018.2807384
7. Han SW, Park SH, Kim HS, Jo MG, Cha HY (2017) Normally-off AlGaN/GaN-on-Si MOS-HFET with a monolithically integrated single-stage inverter as a gate driver. Electron Lett 53:198–199. https://doi.org/10.1049/el.2016.2813, https://ieeexplore.ieee.org/document/7843829/
8. Helmy A, Ismail M (2006) The CHIP – a design guide for reducing substrate noise coupling in RF applications. IEEE Circuits Devices Mag 22(5):7–21. https://doi.org/10.1109/MCD.2006.272996
9. Jenkins K (2004) Substrate coupling noise issues in silicon technology. In: Digest of papers. 2004 Topical meeting on silicon monolithic integrated circuits in RF systems, 2004, pp 91–94. https://doi.org/10.1109/SMIC.2004.1398175
10. Ke X, Sankman J, Song MK, Forghani P, Ma DB (2016) A 3-to-40V 10-to-30MHz automotive-use GaN driver with active BST balancing and VSW dual-edge dead-time modulation achieving 8.3% efficiency improvement and 3.4ns constant propagation delay. In: 2016 IEEE international solid-state circuits conference (ISSCC), pp 302–304. https://doi.org/10.1109/ISSCC.2016.7418027
11. Ke X, Sankman J, Chen Y, He L, Ma DB (2018) A tri-slope gate driving GaN DC–DC converter with spurious noise compression and ringing suppression for automotive applications. IEEE J Solid-State Circuits 53(1):247–260. https://doi.org/10.1109/JSSC.2017.2749041

12. Lidow A, Reusch D, Glaser J (2016) Getting from 48V to load voltage: improving low voltage DC-DC converter performance with GaN transistors. http://www.epc-co.com, Professional Education Seminar at Applied Power Electronics Conference and Exposition APEC 2016

13. Mehrotra V, Arias A, Neft C, Bergman J, Urteaga M, Brar B (2016) GaN HEMT-based>1-GHz speed low-side gate driver and switch monolithic process for 865-MHz power conversion applications. IEEE J Emerg Sel Topics Power Electron 4:918–925. https://doi.org/10.1109/JESTPE.2016.2564946

14. Moench S, Kallfass I, Reiner R, Weiss B, Waltereit P, Quay R, Ambacher O (2016) Single-input GaN gate driver based on depletion-mode logic integrated with a 600V GaN-on-Si power transistor. In: 2016 IEEE 4th workshop on wide bandgap power devices and applications (WiPDA), pp 204–209. https://doi.org/10.1109/WiPDA.2016.7799938

15. Moench S, Hillenbrand P, Hengel P, Kallfass I (2017) Pulsed measurement of sub-nanosecond 1000V/ns switching 600V GaN HEMTs using 1.5GHz low-impedance voltage probe and 50 Ohm scope. In: 2017 IEEE 5th workshop on wide bandgap power devices and applications (WiPDA), pp 132–137. https://doi.org/10.1109/WiPDA.2017.8170535

16. Pfost M, Brenner P, Huttner T, Romanyuk A (2003) A comprehensive experimental study on technology options for reduced substrate coupling in RF and high-speed bipolar circuits. In: Proceedings of the bipolar/BiCMOS circuits and technology meeting, 2003, pp 39–42. https://doi.org/10.1109/BIPOL.2003.1274931

17. Reiner R, Waltereit P, Weiss B, Moench S, Wespel M, Müller S, Quay R, Ambacher O (2018) Monolithically integrated power circuits in high-voltage GaN-on-Si heterojunction technology. IET Power Electron 11:681–688. https://doi.org/10.1049/iet-pel.2017.0397, https://ieeexplore.ieee.org/document/8338223/

18. Schroter P, Jahn S, Klotz F (2011) Improving the immunity of automotive ICs by controlling RF substrate coupling. In: 2011 8th workshop on electromagnetic compatibility of integrated circuits, pp 182–187

19. Su D, Loinaz M, Masui S, Wooley B (1993) Experimental results and modeling techniques for substrate noise in mixed-signal integrated circuits. IEEE J Solid-State Circuits 28(4):420–430. https://doi.org/10.1109/4.210024

20. Zhang Y, Rodríguez M, Maksimović D (2016) Very high frequency PWM buck converters using monolithic GaN half-bridge power stages with integrated gate drivers. IEEE Trans Power Electron 31(11):7926–7942. https://doi.org/10.1109/TPEL.2015.2513058

Chapter 4
Design of Fast-Switching Circuit Blocks

This chapter presents the design of the fast-switching circuit blocks. For the output voltage regulation of fast-switching converters, a conventional voltage mode control (VMC) is suitable, introduced in Sect. 4.1. The critical sub-circuits are the blocks generating or propagating the switching signals, which are the PWM generator including an error voltage comparator, level shifter, and gate driver. This section proposes a design of these circuit blocks, which meet the requirements to generate and propagate PWM on-time pulses shorter than 3 ns. As demonstrated in Sect. 2.3.6, this allows an operation at switching frequencies as high as 30 MHz at input voltages of 50 V or higher.

4.1 Output Voltage Regulation

The voltage-mode control used for the output voltage regulation of the buck converters covered in this book is depicted in Fig. 4.1 for the example of a synchronous buck converter. An error amplifier compares the divided output voltage to a reference voltage V_{ref} to generate an error signal V_{err}. A PWM comparator compares V_{err} to a sawtooth signal, and thus creates a PWM signal. The PWM signal is fed to a dead time control block, which generates the non-overlapping control signal $CTRL_{hs}$ and $CTRL_{ls}$ for the low-side and high-side switch MN_{LS} and MN_{HS}. $CTRL_{hs}$ is passed through a level shifter to a gate driver on the high-side domain to control the high-side switch. $CTRL_{ls}$ directly connects to the low-side gate driver, controlling the low-side switch. The duty cycle of the PWM signal controls the output voltage. In an asynchronous converter, no dead time control and no low-side gate driver are required (compare Fig. 3.1a). A conventional type III compensation is applied to achieve a stable regulation loop [2].

An increase of the switching frequency along with higher conversion ratios affects all switching blocks. Besides handling a higher switching frequency, the

© Springer Nature Switzerland AG 2020

J. Wittmann, *Integrated High-V*$_{in}$ *Multi-MHz Converters*,

https://doi.org/10.1007/978-3-030-25257-1_4

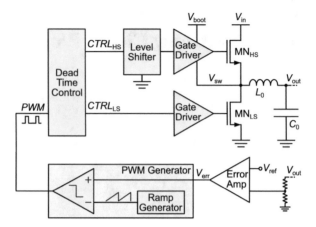

Fig. 4.1 Voltage mode control (VMC) in a synchronous buck converter

main limitation is the generation and propagation of the minimum required on-time pulse as low as 3 ns at low PWM duty cycles. This short pulse has to be generated by the PWM generator and propagated safely through the dead time control, level shifter, and gate driver. The gate driver has to be strong enough to provide this small pulse even at gates of the power switches with a large capacitive load.

A current-mode control (CMC) [2] is often the preferred as it is robust against input voltage variations, it provides a better transient behavior in case of load steps, and it comes along with a current limitation functionality. However, it requires a current sensor, which generates the PWM generator ramp out of the inductor current ramp. The settling time of the current sensor limits the minimum possible on-time of the converter in peak control mode, which senses the inductor current at the high-side switch. For example, to generate on-time pulses of a few nanoseconds at multi-MHz switching, the current sensor has to settle also in this time frame, when it gets activated as soon as the high-side switch is turned on. For high conversion ratios, this can be overcome by switching to a valley control by sensing the inductor current at the low-side switch. The resulting ramp controls the off-time of the PWM signal. However, valley control limits minimum possible off-time of the converter, and thus does not allow an operation at low conversion ratio. A widely varying input voltage at switching frequencies in the multi-MHz range is thus limited by the current sensor.

A voltage-mode control is chosen for the fast-switching buck converters of this book to overcome the limitation by the current sensor, as both a wide input voltage range and switching frequency up to 30 MHz are the target operating region. Rather than the ramp generator, a current-mode control with an optimized current sensor (not covered in this book) would require the same fast-switching circuit blocks to transfer the PWM signal to the power switches.

The transient response on line or load steps in a voltage-mode control significantly benefits from a higher switching frequency, and thus is able to cope with transient regulation performance achieved with a current-mode control. In a voltage-mode control, the transient response improves with increasing the crossover frequency f_{sw}. To avoid an interference, f_{sw} should be at least four times lower than the switching frequency [2]. As a second constraint, f_{sw} has to be higher, than the resonant frequency of the output filter to avoid instabilities. Increasing the switching frequency, in order to decrease the output filter, increases the output filter resonant frequency, and thus makes a faster regulation even mandatory.

A design of each of the fast-switching circuit blocks in the voltage-mode control loop, highlighted in Fig. 4.1, is proposed subsequently, which fulfills the requirement to operate the converter with PWM on-time pulses down to the range of 3 ns.

4.2 Sawtooth and PWM Generator

The PWM generation proposed for the fast-switching converters, covered in this book, is based on a sawtooth signal. A sawtooth signal is often preferred, as the rising and falling PWM edge can be derived from a different signal, e.g., the oscillator clock. This avoids that glitches cause a multiple turn-high and turn-low at the PWM signal, resulting in a multiple switching of the power switches. Alternatively, a triangular signal could be used to be compared with the error signal, as it is typically not preferred due to the sensitivity to signals, as both PWM edges have to be derived from a single signal. However, some of the proposed techniques in this section are also appropriate for a triangular implementation.

4.2.1 Requirements for Sawtooth and PWM Generator

The signals of a basic sawtooth based PWM generator are shown in Fig. 4.2. The sawtooth ramps are started at a voltage $V_{ref,lo}$ which is above the minimum output voltage of the error amplifier, while the maximum ramp voltage $V_{ref,hi}$ is below the maximum output voltage of the amplifier (Fig. 4.2). Thus, the error signal from the error amplifier V_{err} can fall below and rise above the signal range of the sawtooth ramps, which allows a duty cycle range from 0% to 100%.

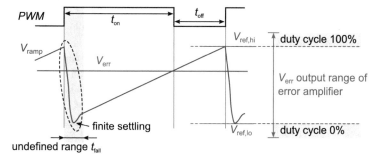

Fig. 4.2 PWM generation based on a conventional ramp signal

The minimum duty cycle is limited by the falling slope of the sawtooth signal. As shown in Fig. 4.2, the ramp of the sawtooth signal is typically reset at the rising edge of the PWM signal. This requires a finite settling time t_{fall} of the rising slope. Thus, the minimum pulse width of the PWM signal has to be limited to $t_{on} > t_{fall}$, as the input signal of the PWM comparator is not valid during t_{fall} and the PWM signal cannot be set low correctly.

As the challenging timing requirements are difficult to fulfill by the control circuits, the converter can be switched to constant on-time mode [4]. The on-time is kept constant at a feasible value, e.g., 5 ns, while the off-time is adjusted, to regulate the output voltage. The drawback is a widely varying switching frequency from the operating frequency down to a few MHz.

Converters suitable for input voltages as large as 50 V and above require a high-voltage technology for the power stage. These technologies are typically not optimized for speed and high bandwidth. Even low-voltage transistors have moderate bandwidths in the range of 1 GHz and limit the speed of analog circuits to the nanosecond range.

While level shifters and gate drivers can propagate on-time pulses of less than 3 ns [18], a sawtooth generator is required, which is suitable for PWM signals with on-time pulses in the same range. Therefore, the focus of this book is on sawtooth generation for PWM signals with on-time pulses of <2 ns.

4.2.2   Limitations of Conventional Sawtooth Generators

Conventional concepts to generate sawtooth signals are shown in Fig. 4.3 [1, 3, 16]. Figure 4.3a depicts a concept, in which a linear regulator generates the lower sawtooth level reference $V_{ref,lo}$. A switch connects a capacitor to $V_{ref,lo}$. A constant reference current I_{ramp} is charging up the capacitance and thus generates the sawtooth ramp. A comparator detects, when the ramp reaches the upper level reference $V_{ref,hi}$, and the ramp is reset again to $V_{ref,lo}$ by turning on the switch for a short time. The falling slope, shown in Fig. 4.3a, is limited by the on-resistance of the switch. A larger switch would introduce parasitic coupling and add a non-linear parasitic capacitance to the main capacitor. During the falling slope, either the maximum on- or the maximum off-time of the PWM signal is limited, depending on the time when PWM signal is reset. A large discharge current to the output of the linear regulator would cause a significant voltage overshoot at $V_{ref,lo}$. A large buffer on-chip capacitance would reduce the overshoot. However, on-chip capacitances are still too small to fully eliminate the overshoot. The larger capacitances would make the linear regulator even slower to settle back to the nominal voltage.

The coupling to the reference can be avoided with the concept shown in Fig. 4.3b. A capacitor is charged and discharged with an identical reference current I_{ramp} resulting in a triangular signal, which is undesired as both the rising and falling PWM edge are derived from the same signal, as previously described. The upper and the lower level $V_{ref,hi}$ and $V_{ref,lo}$ of the sawtooth signal are detected by a comparator.

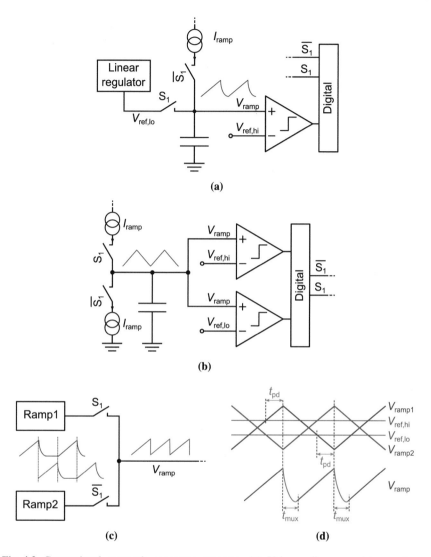

Fig. 4.3 Conventional sawtooth generator concepts. (**a**) Using a linear regulator as lower reference. (**b**) Sawtooth generation based on a triangular signal generator. (**c**) Multiplexing two interleaved sawtooth signals. (**d**) Resulting sawtooth signal generated by two multiplexed triangular signal generators

For both Fig. 4.3a and b, a faster falling slope is achieved by multiplexing two phase-shifted sawtooth signals, as shown in Fig. 4.3c. Only the parasitic capacitances of the switches in the multiplexer and the output node have to be discharged. Two identical triangular generators are required to be multiplexed to obtain a sawtooth signal. The resulting signals of the multiplexed triangular signals and the resulting sawtooth signal are shown in Fig. 4.3d. At high frequencies, the ramp-up time, and thus the period of the multiplexed sawtooth signal, is mainly determined by the propagation delay t_{pd} of the comparator, which varies over process variations and temperature by more than 30% [15]. This drawback also applies to the single stage generator of Fig. 4.3b.

An additional disadvantage of the multiplexed sawtooth concept is that during the fast falling slope at the output of the multiplexer (see Fig. 4.3c) the input capacitances of the following PWM comparator and the capacitances of the multiplexer have to be discharged. As shown in Fig. 4.3d, the falling slope during t_{mux} also significantly limits the minimum on-time of the PWM signal [17]. A decrease of the resistance of the multiplexer would result in the same limited falling slope, as the parasitic capacitances of the multiplexer would increase proportionally.

4.2.2.1   Proposed Sawtooth Based PWM Generator

The proposed configurable sawtooth based PWM generator, shown in Fig. 4.4, is able to generate PWM pulses as small as 2 ns. The reference current I_{ramp} and the upper ramp reference $V_{ref,hi}$ are adjustable and thus allow to control the slope and amplitude of the ramps. This enables a wide frequency range from <2 MHz up to even >50 MHz.

Two parallel integrator stages are used as sawtooth generators to obtain inter-leaved ramp signals V_{ramp1} and V_{ramp2} [6]. Each stage uses two comparators (PWM and CLK) directly at the integrator output. The signals of the proposed sawtooth generator are shown in Fig. 4.5. Both CLK comparators detect when the ramp reaches the upper sawtooth level $V_{ref,hi}$. When a CLK comparator detects a ramp crossing $V_{ref,hi}$, the ramp is reset and the ramp of the complementary stage is

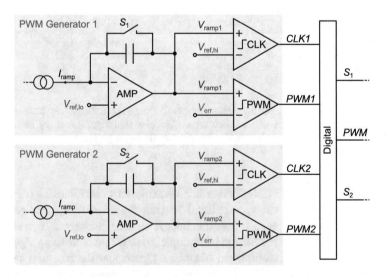

Fig. 4.4 Proposed PWM generator, based on two interleaved integrator stages

Fig. 4.5 Signals of the PWM
generator

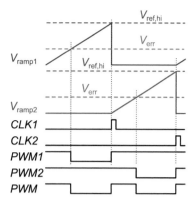

started. This determines the switching frequency of the converter (Fig. 4.5). Each
of the PWM comparators at the integrator outputs generates the signals $PWM1$ and
$PWM2$ during the ramp phase of its stage by comparing the ramp to the voltage
V_{err} of the error amplifier. The final PWM signal is generated by a digital circuit.
The PWM signal is set high when any of V_{ramp1} or V_{ramp2} crosses $V_{ref,hi}$, and is set
low on each of the falling edges of $PWM1$ and $PWM2$. This way, both signals
$PWM1$ and $PWM2$ are multiplexed in the digital domain.

The ramps of the sawtooth signals are generated by an integrator. The output of
an amplifier is regulated to the lower reference $V_{ref,lo}$, while the switches S_1 and S_2
short the feedback capacitor. When S_1 or S_2 is turned off, the current I_{ramp}, drawn
out of the negative input of the amplifier, is integrated, and a linear ramp is obtained.
The reset of each ramp can occur slowly during the ramp time of the complementary
integrator stage.

A standard symmetrical amplifier is used for the integrator, which is suitable
to achieve the required bandwidth. It is fast enough to settle during the ramp
time of the complementary integrator stage. The amplifier in the integrator can
be designed faster than a linear regulator used for the concept of Fig. 4.3a. As the
feedback capacitor of the integrator provides a Miller-compensation at the same
time, no additional capacitance has to be added to stabilize the integrator. The switch
across the feedback capacitor is implemented as a transmission gate. Both parallel
PMOS and NMOS switches have identical parasitic capacitances to compensate the
coupled charge during their complementary switching. While conventional concepts
have closed switches in the charging path, the switches in the proposed concept are
open during the ramp-up. Thus, the resistance and the size of the switches can be
kept small, and the influence on the ramp is negligible. The switches only have
to ensure that the sawtooth signal is discharged during the charging time of the
complementary ramp.

All four comparators are implemented as 3-stage comparators to ensure a fast
detection of a very small voltage difference at the comparator input. The first
and second stage are designed fully differential with a sufficient gain to increase
the voltage swing. In the third stage, which provides a single-ended output with
full voltage swing for the digital circuit, the charging and discharging of parasitic

capacitances are dominant. Thus, instead of maximizing the gain, the sizes of all transistors in the signal path are kept to a minimum. A further buffer stage is used to drive the digital load.

Compared to the conventional concepts in Fig. 4.3, the PWM comparator has to be used two times, but no multiplexing of the sawtooth ramps is required. As the CLK and PWM comparators are independent of each other, the minimum on-time of the PWM signal only depends on the propagation delay of the PWM comparator and the digital circuit, and is not limited to the time t_{mux} of the multiplexed falling edge of the sawtooth signal using an analog multiplexer, as shown in Fig. 4.3d. Another advantage of this concept is that the charging current of the capacitor is not drawn from the reference $V_{ref,lo}$, as it is the case in the concept shown in Fig. 4.3a. A common reference can be used for all integrator stages without any interference.

The proposed PWM generator was implemented in the 180 nm high-voltage BiCMOS technology used for the converters of this book. Measurement results of the sawtooth generator at a frequency of 10 MHz are depicted in Fig. 4.6a. The interleaved ramps are started at $V_{ref,lo} = 1.5$ V. The ramp is charged with $I_{ramp} = 26\,\mu A$ up to the upper reference $V_{ref,hi} = 2.8$ V. The reference V_{err} for both PWM comparators is set in the center between $V_{ref,lo}$ and $V_{ref,hi}$ to $V_{err} = 2.15$ V. When the ramp reaches $V_{ref,hi}$, the second ramp V_{ramp2} is started and the PWM signal is set to high after t_1. When V_{ramp2} reaches V_{err}, the PWM comparator turns low and sets the falling edge of the PWM signal with a delay of the comparator and the digital circuit after t_2. The minimum possible on-time of PWM would be limited to t_2 if V_{err} is very close to $V_{ref,lo}$, and the PWM comparator switches immediately after the ramp has started. To reduce the minimum possible on-time to a minimum, an additional digital delay of approximately 3 ns is added to t_1 to delay the rising edge of the on-time pulse, and thus compensate for the comparator delay.

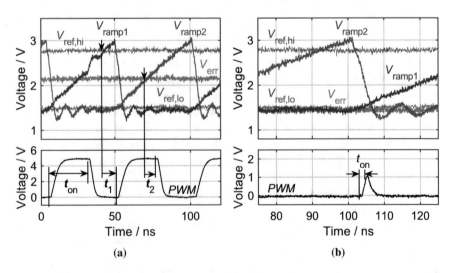

Fig. 4.6 Measured sawtooth generator signals at 10 MHz and (a) 50% PWM duty cycle and (b) minimum PWM on-time of 2 ns

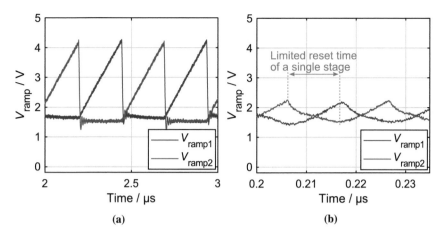

Fig. 4.7 Complementary sawtooth signals measured at (**a**) single stage frequency of 2 MHz (resulting PWM frequency: 4 MHz) (**b**) single stage frequency of 50 MHz (resulting PWM frequency: 100 MHz)

The resulting duty cycle of $D = 50.5\%$ matches the expected value. A minimum on-time of the PWM signal of $t_{on} = 2\,ns$ is achieved in a measurement shown in Fig. 4.6b. Due to the interleaved stages, an operation frequency of 10 MHz in each stage results in a PWM frequency of 20 MHz at the output. The slopes of the digital nodes, i.e. the comparator outputs and the PWM signal at the digital output, are significantly reduced in the on-chip measurement, as the probes added a considerable capacitance. Even with Picoprobes, the probe capacitance is at least 5 times larger than the node capacitances. Thus, at minimum on-time, a rail-to-rail switching of the PWM signal cannot be measured, but the turn-high and turn-low events are indicated.

The frequency of the sawtooth stages can be adjusted over a wide range by either adjusting the upper reference $V_{ref,hi}$ or by a variation of the externally adjustable reference current $I_{ramp1} = I_{ramp2}$. Figure 4.7 shows measurements at a sawtooth frequency of 2 MHz and 50 MHz, which results in a PWM frequency of 4 MHz and 100 MHz, respectively. Thus, a converter with a switching frequency of 100 MHz could still work with a duty cycle down to 20% using the proposed sawtooth based PWM generator.

4.3   Level Shifters

As demonstrated in Sect. 3.1, the requirements for a level shifter controlling a PMOS high-side switch (subsequently called PMOS level shifter) differ from requirements for a level shifter controlling an NMOS high-side switch (subsequently called NMOS level shifter). While the high-side gate driver supply for a PMOS switch is practically static, the high-side gate driver supply for an NMOS is floating. It suffers from fast-switching transitions across the entire V_{in} range, as its ground is referenced to the switching node V_{sw}.

This section elaborates the limitations of conventional level shifter concepts [9–14]. It shows first a PMOS level shifter design, which is optimized for short propagation delays , resulting in a capability to transfer the required minimum on-time pulses as small as 3 ns suitable for high-V_{in} multi-MHz converter. Secondly, an NMOS level shifter is proposed which is additionally very robust against coupling, caused by fast-switching transitions.

4.3.1   Conventional Level Shifter Concepts

Several fast-switching NMOS and PMOS level shifters have been published [9, 12–14], which achieve a very fast propagation delay, but a transfer of short pulses is not supported. The signal is typically transferred by a high-voltage or cascaded switch on the low-side, which creates a voltage drop on the high-side across a resistor, or a current source propagates the signal current, as shown in Fig. 4.8. This voltage drop is detected by a digital circuit or latch on the high-side [12]. As the high-side circuits are using low-voltage transistors, the voltage drop across a resistor or current source has to be clamped to not violate the maximum ratings. A diode (Fig. 4.8a) would limit the voltage at the logic input (node D) to one diode drop below $HSGND$. As the maximum voltage rating of the logic is not significantly higher than the supply voltage of the high-side, the diode requires to have low series resistance to achieve a low forward voltage drop. Hence, the diode has to be very large in area, which results in a large parasitic capacitance. Alternatively, the current through the low-side high-voltage switch could be limited. In both cases, the speed is too low for small on-time pulses.

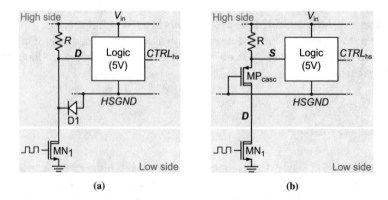

Fig. 4.8 Conventional level shifter concepts with (**a**) diode clamping and (**b**) PMOS cascode

A second option is to use a high-voltage PMOS cascode (Fig. 4.8b), which clamp the voltage slightly above $HSGND$, but the PMOS cascode MP_{casc} limits also the charging current, as it cuts off at falling voltages at node S. With a large switch current, S and D could still be discharged quickly, but the next logic transition is only possible after D and S are fully charged back to V_{in} via the resistor R. Simulations confirmed that large parasitic capacitances of the high-voltage transistors lead to a charging time in the range of 30–100 ns in the 180 nm high-voltage BiCMOS technology. Pulses within this time are completely filtered. As the switching period at frequencies higher than 10 MHz is shorter than the filter time, the conventional concepts of Fig. 4.8 are not suitable. Capacitive level shifters are fast, but large coupling currents occur during high-side transitions over the capacitances [9].

4.3.2 Level Shifter for PMOS High-Side Switches

A PMOS level shifter design suitable for fast-switching high-V_{in} converters is shown in Fig. 4.9. The level shifter is based on a symmetrical single-stage amplifier. Its input pair is controlled by the PWM and inverted PWM signal on the low-side, while the single-ended output of the amplifier $CTRL_{hs}$, connecting to the gate driver, is referenced to the high-side supply. This concept has been previously proven to achieve very short propagation delays in sub-nanosecond range in [7], which uses a

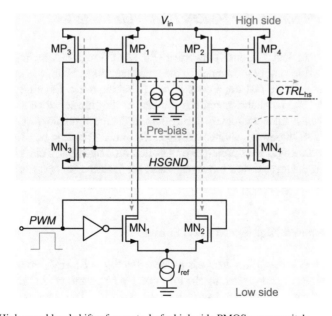

Fig. 4.9 High-speed level shifter for control of a high-side PMOS power switch

latch at the high-side current mirror input to achieve a fast switching and thus short propagation delays. However, a latch at the high-side current mirror inputs requires strong low-side switches to change the latching state of the high-side input.

The pre-bias current sources are connected to the high-side current mirror inputs of the amplifier (MP_1 and MP_2). Pre-bias currents in the nanoampere range maintain the current mirrors input voltages MP_1 and MP_2 close to their threshold voltages. This way, only a very small voltage swing at the current mirror inputs is required to activate one of both current mirrors if PWM changes it's state. Due to the small voltage swing, a sub-nanosecond propagation delay is achieved. The proposed pre-biasing allows to use smaller high-voltage low-side switches, as the signal currents through the low-side switches are immediately transferred to the output, as the current mirrors are kept always at their conduction boundary. Compared to a latched level shifter, a static current is required to keep the level shifter output in its current state. However, the required current is not significant, compared to a fast switching power stage in high-V_{in} multi-MHz operation.

The presented PMOS level shifter is used in the asynchronous buck converter, simulated for efficiency comparison in Sect. 5.6, as well as in the resonant stage of the proposed parallel-resonant converter (PRC) presented in Sect. 7.2. Experimental results of the PRC confirm a proper operation of the PMOS level shifter across a wide input-voltage range up to 48 V at switching frequencies up to 25 MHz. The PRC also confirms a proper transfer of on-time pulses down to 3 ns to the high-side, as they are required for the PRC to operate across the wide input voltage range.

4.3.3 Level Shifter for NMOS High-Side Switches

As introduced in Sect. 3.1, buck converters with NMOS high-side switches require a floating high-side supply referenced to the switching node. Charge is coupling into the signal path during fast transitions of the switching node. Existing level shifter concepts for PMOS switches cannot be utilized, as they are sensitive to in-coupling and would cause false switching during fast high-side transitions. Level shifters suitable for NMOS power switches (NMOS level shifters) have to be very robust and require protection against coupling during switching transitions. This is in particular challenging if the level shifter has a very short propagation delay on both switching edges, in order to propagate on-time pulses as short as 3 ns.

4.3.3.1 Proposed High-Speed Level Shifter

A robust NMOS level shifter design is proposed in Fig. 4.10. High-voltage switches (MN_0 and MN_1) are controlled by the low-side PWM signal, while MN_0 and MN_1 are connected each to a current source load I_{up} from V_{in}. This creates a differential voltage ΔV_{sig} at nodes A and B, which is detected by a high-speed comparator to provide the high-side PWM signal $CTRL_{hs}$ between V_{boot} and the high-side

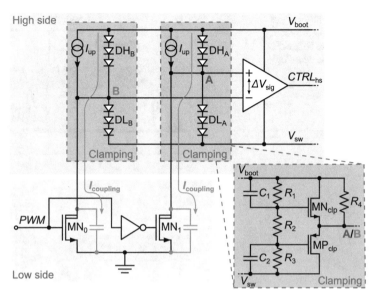

Fig. 4.10 High-speed level shifter for control of NMOS power switches at a floating high-side

ground, which is the switching node voltage V_{sw}. The differential voltage range at the nodes A and B is limited by an overlapping clamping, implemented by diodes, as shown in Fig. 4.11a. At each node A and B, a diode stack connected to V_{boot} limits the voltage to a lower boundary. Another diode stack connected to V_{sw} limits the voltage to an upper boundary. When the low-side switch MN_0 or MN_1 is turned on at time t_1 (Fig. 4.11a), the according node A or B is held at the lower boundary by the clamps DH_A or DH_B, respectively. Accordingly, when the low-side switch is turned off, the clamp DL_A or DL_B holds the voltage at A or B at the upper voltage boundary, while the clamp conducts the current I_{up} of the current source connected to V_{boot}.

The clamping voltages are designed such that none of the diode stacks is conducting in between the upper and lower voltage boundary. A high-impedance region is obtained, in which the differential PWM signal at A and B can change its state very fast at time t_1 (Fig. 4.11a). After the state change of the PWM signal, the NMOS power switch is turned on or off by the gate driver at time t_2. A high-side transition follows, in which the voltage rails V_{sw} and V_{boot} experience a voltage step up to 50 V with a slope of tens of V/ns.

The parasitic capacitances at the drain of MN_0 and MN_1 are charged. Even for small parasitic capacitances, coupling currents $I_{coupling}$ in the range of 2 mA are generated, which are superimposed to the signal currents in the range of 50 μA, as demonstrated in Fig. 3.2b. At rising high-side transitions during t_2-t_3, the forward voltages across the clamps DH_A and DH_B cause the voltages at nodes A and B to fall significantly below the lower boundary of the high-impedance region. The overlapping diode clamps are designed such that the high-impedance region

Fig. 4.11 Level shifter signal
at the high-side comparator
input: (**a**) Transient behavior;
(**b**) influence of the coupling
currents

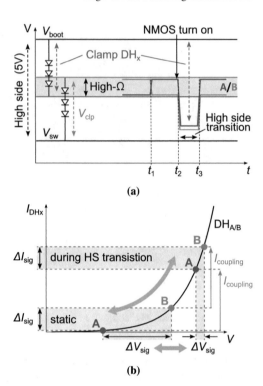

typically has a width of a few hundred mV and occurs in the center between V_{boot}
and V_{sw}. This allows the nodes A and B to drop by nearly two volts below the
high-impedance region. With a large common mode input range of the comparator,
a very large forward voltage across the clamps can be tolerated. Thus, the clamps
can be designed small to handle the full coupling currents. This is a major advantage
compared to the conventional clamps of Fig. 4.8 with large area.

At falling high-side transitions, the coupling currents are clamped by DL_A and
DL_B to V_{sw}, accordingly, such that the nodes A and B are pulled up above the upper
boundary of the high-impedance region.

4.3.3.2 Robustness Considerations

During high-side transitions, the level shifter does not have any speed requirement,
as the PWM signal has already been transferred to the high-side switch. However,
the level shifter needs to keep the signal state to avoid a false switching of the level
shifter output and thus of the power switch. The robustness is analyzed in Fig. 4.11b
by means of the diode VI characteristics of the clamps. In the static state, before and
after the high-side transition, either the low-side switch MN_0 or MN_1 is turned on.
The switch current is flowing through one of the diode stacks DH_A or DH_B, while
the current through the other diode stack is zero, resulting in a particular differential

voltage ΔV_{sig} between nodes A and B. ΔV_{sig} is detected by the comparator. Node A in Fig. 4.11b corresponds to a current in the diode clamps, which is close to zero (assuming that MN_0 is turned on) resulting in a very low voltage on the VI curve, while node B is shifted by the signal current to a higher voltage level. During the high-side transitions ($t_2 - t_3$ in Fig. 4.11a), a coupling current $I_{coupling}$ is added to the signal current. As depicted in Fig. 4.11b, both nodes A and B are shifted to higher current levels on the diode curves by the same coupling current $I_{coupling}$ due to the symmetry (same size of MN_0 and MN_1). The current difference on the VI curve between the nodes A and B remains unchanged and is equal to the signal current ΔI_{sig}. With the proposed concept, the voltage difference of ΔV_{sig} is reduced during high-side transitions, but it remains positive, even at large coupling currents. As the signal ΔV_{sig} does not change polarity, the state of the comparator is not affected by coupling currents, and a high level of robustness of the level shifter is achieved.

4.3.3.3 Implementation of the Clamping Structures

A further improved implementation of the clamp circuits is proposed in Fig. 4.10 (subset). The diode stacks are replaced by an NMOS transistor connected to V_{boot}, and a PMOS transistor connected to V_{sw}. Fixed gate bias voltages are provided by a resistor divider (R_1, R_2, and R_3) such that a matching clamping behavior for the upper and lower clamping level is achieved, and the transistors MN_{clp} and MP_{clp} are conducting if the voltage at node A or B is below or above the high-impedance region, shown in Fig. 4.11a. Small buffer capacitors (C_1 and C_2) are required to keep the gate voltages constant during fast signal transients, which are coupling to the gates. Compared to real diodes, the biased transistors can conduct much higher currents during clamping of the coupling. The implementation is comparable to a class AB output stage. At the same time, the voltage swing at the comparator inputs during high-side transitions can be further limited. A higher margin to the limits of the common mode input range of the comparator is achieved, or higher coupling currents can be allowed, which further increases the robustness of the level shifter.

Experimental results of the level shifter, implemented in a fast-switching buck converter are shown in Sect. 4.5.

4.4 Gate Driver

The gate driver has to be able to charge and discharge the gate capacitance of the high-side switch MN_{HS} or low-side switch MN_{LS} of the buck converter (Fig. 4.1) in the time range of ≈ 1 ns, to achieve PWM on-time pulse down to 3 ns (see Sect. 2.3.6.2). As fast transitions at high input voltages cause large coupling currents into the gate through the drain-gate capacitances of the NMOS transistors, the gate driver also has to be able to dissipate these currents, while it keeps the NMOS transistors in its current switching state. Furthermore, the propagation delay of the

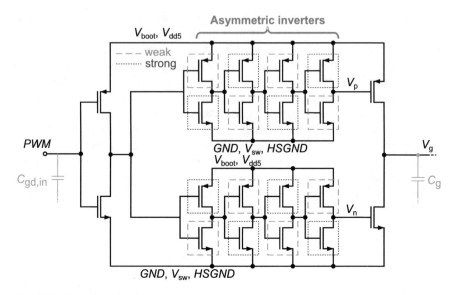

Fig. 4.12 Gate driver circuit

gate driver has to be optimized to be within a few nanoseconds, and the power consumption should be minimal. This is achieved by a tapered buffer, consisting of several inverter stages in series, as shown in Fig. 4.12. The delay is optimal if the driver strength is increased from stage to stage by a factor of α [8]. The factor α is determined by the ratio of the capacitance at the input of the first inverter stage $C_{gd,in}$ to the gate capacitance C_g of the NMOS switch (MN$_{LS}$ or MN$_{HS}$) with $\alpha^N \cdot C_g/C_{gd,in}$, where N is the number of inverter stages. As the driver strength of the gate driver output stage is depending on the charge to be delivered to the power switch, C_g represents the effective capacitance at the gate, rather than the physical capacitance, as the switching node transition couples back a large amount of charge to the gate over the gate-drain capacitance C_{gd}. Thus, the effective capacitance is determined by including the Miller capacitance of C_{gd} in the total capacitance of C_g. Simulations have shown that this approach resulted in a Miller plateau at the gate, which is sufficiently below the threshold in off-state, and close to the gate driver supply during on-state. This allows to assume that the power switch is already fully turned-on during the Miller plateau. A compromise between die size and propagation delay was found for $N = 6$ inverter stages with $\alpha = 5.24$, $C_{gd,in} = 7\,\text{pF}$ and $C_g = 145\,\text{pF}$. To improve the efficiency, cross-currents have to be avoided, in particular in the last driver stage. This is accomplished by splitting up the driver stage into two branches [5]. This allows to control the PMOS and NMOS transistor of the last driver stage separately. There is no area penalty, since each branch has to drive approximately half of the capacitive load. A protection against cross-currents in the last stage is achieved by an asymmetry factor in the inverter stages as indicated in Fig. 4.12 [5]. By alternating the NMOS and PMOS transistor strengths from stage to stage from weak to strong in both branches, the rising PWM signal is delayed in the NMOS branch, while the falling PWM signal is delayed in the PMOS branch. This assures that the PMOS is always turned off before the NMOS turns on, and vice versa. An asymmetry factor of 20% is used in this design, i.e. in the inverters of each branch, the width of the strong transistors is increased and the width of the weak transistors is decreased by 20% with respect to the nominal value. The asymmetric design reduces the power consumption of the gate driver by more than 23%, verified by simulation. With a total power dissipation of 40 mW at 10 MHz, even the asymmetric gate driver contributes significantly to the overall converter losses.

4.5 Experimental Results of Fast-Switching Circuit Blocks

The fast-switching circuit blocks were implemented as part of a configurable voltage converter (180 nm high-voltage BiCMOS technology). The voltage converter contains two power stages with once a PMOS power switch controlled by the presented PMOS level shifter (see Sect. 4.3.2) and once with an NMOS power switch, which is controlled by the proposed NMOS level shifter (see Sect. 4.3.3). Both include the gate driver described in Sect. 4.4, which was sized identically for each PMOS and NMOS power stage. This allows a configuration as asynchronous buck converter with either a PMOS high-side switch or an NMOS high-side switch. Furthermore, the converter can be configured as quasi-resonant converter (QRC), which is covered in Sect. 7.1, as well as a parallel-resonant converter (PRC), which covered in Sect. 7.2 later in this book.

A micro-photograph of the fast-switching circuit blocks as part of the configurable voltage converter is shown in Fig. 4.13. It contains the PWM generator with a die size of $0.11\,\mathrm{mm}^2$, the PMOS level shifter with a die size of $0.03\,\mathrm{mm}^2$, the

Fig. 4.13 Micro-photograph of a configurable voltage converter implementation, including an NMOS and PMOS power stage, suitable for configuration each as asynchronous buck converters, as quasi-resonant converter (see Sect. 7.1), and as parallel-resonant converter (see Sect. 7.2)

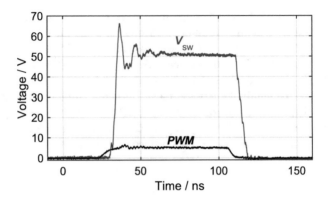

Fig. 4.14 Experimental verification of the level shifter. Correct switching of the buck converter at an input voltage of 50 V

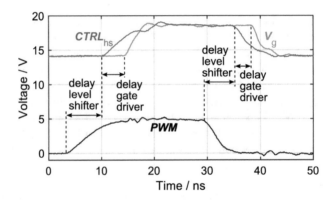

Fig. 4.15 Signals of the internal nodes of the NMOS power stage

NMOS level shifter with each a die size of $0.04\,\text{mm}^2$, and the gate drivers with a die size of $0.08\,\text{mm}^2$ each. The NMOS level shifter and the according gate driver are placed in the high-side isolation well (as depicted in Sect. 3.4) to separate the floating high-side supply from the substrate in order to control the NMOS power switch.

Subsequently, experimental results of the implemented NMOS power stage are shown (with the converter configured as asynchronous buck converter). In Fig. 4.14, the input of the NMOS level shifter (PWM) and the switching node V_{sw} were measured in buck operation at a high input voltage of $V_{\text{in}} = 50\,\text{V}$ to verify the robustness of the level shifter. The results prove that no false switching occurs, even at fast rising and falling slopes of the high-side of up to $20\,\text{V/ns}$ (simulations even confirmed a robust operation up to $80\,\text{V/ns}$). The current consumption of the level shifter is in the range of $300\,\mu\text{A}$.

The results of a propagation delay measurement of the PWM signal transitions are shown in Fig. 4.15. Low-capacitive Picoprobes are used to directly measure

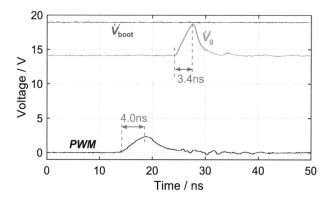

Fig. 4.16 Minimum on-time pulse of the level shifter

the signal on-chip. Due to the limited maximum voltage rating of the probes, the switching had to be kept constant at $V_{sw} = 14$ V, and each probe was referenced to either the high-side or the low-side supply. However, the propagation delay of the power is independent of the power switch switching, as it occurs only after the PWM signal propagated to the gate V_g. A maximum delay of 5 ns is measured for both the rising and the falling PWM signal edge from the level shifter input (PWM) to the level shifter output ($CTRL_{hs}$). Due to the high symmetry of the level shifter, the resulting on-time at the level shifter output matches the on-time of the PWM signal provided at the level shifter input at the low-side with a high accuracy. The optimized gate driver [17] requires about 3 ns until the gate voltage V_g of the NMOS switch is turned on or off. The propagation delay did not vary significantly with the input voltage V_{in}. The actual propagation delay is expected to be even faster than measured, as the low-capacitive Picoprobes still add a significant load to the minimum sized gates at the measured nodes of the level shifter.

The minimum possible on-time of the PWM signal, which can be transferred to the high-side by the NMOS level shifter and propagated by the gate driver, is measured in Fig. 4.16. The proposed PWM generator provides a minimum on-time pulse of 4 ns to the NMOS level shifter input. The on-time pulse cannot be measured accurately due to the load of the Picoprobes, which significantly flatten the measured pulse; however, the instances, when PWM starts to rise or fall indicate the real pulse without the load of the probe. It can be observed that the minimum on-time pulse propagates correctly to the gate of the NMOS power switch and the gate driver output V_g with a slightly reduced on-time of 3.4 ns (note that the probe on PWM at the low-side was released for the measurement of the pulse at V_g to obtain a correct pulse at the level shifter input). This measurement confirms that pulses in the range of 3 ns can be generated at the gate of the NMOS power switch with a high accuracy of significantly below 1 ns.

It can be observed that the slope at V_g during the gate turn-on and off is limited to about 3 ns in measurement, while simulations (not including parasitic board and bond inductances) showed a slope of 1 ns. The limitation of a proper shaped pulse is limited by the parasitic inductances of the PCB design in the gate current loop (compare Fig. 3.5), which was not yet finally optimized according to Fig. 3.6c in the measurements of this section. The impact of the non-optimized board becomes also visible in the large overshoot at the switching node V_{sw} at the high-side switch turn-on, shown in Fig. 4.14. A further improvement according to Fig. 3.6c, including direct-bond to PCB, becomes essential, and was used for the further implementations of the converters presented in Chaps. 6 and 7.

The experimental results confirm that an operation of the power stage allows very small duty cycles to enable an operation at high input voltages up to 50 V at switching frequencies up to 30 MHz or higher.

References

1. Azais F, Bernard S, Bertrand Y, Michel X, Renovell M (2001) A low-cost adaptive ramp generator for analog BIST applications. In: 19th IEEE proceedings on VLSI test symposium. VTS 2001, pp 266–271. https://doi.org/10.1109/VTS.2001.923449
2. Basso C (2008) Switch-mode power supplies spice simulations and practical designs, 1st edn. McGraw-Hill, Inc., New York
3. De Lima J, Pimenta W (2007) A gm-C ramp generator for voltage feedforward control of DC-DC switching regulators. In: 2007 IEEE international symposium on circuits and systems, pp 1919–1922. https://doi.org/10.1109/ISCAS.2007.378350
4. Fang CC (2012) Closed-form critical conditions of instabilities for constant on-time controlled buck converters. IEEE Trans Circuits Syst I 59(12):3090–3097. https://doi.org/10.1109/TCSI.2012.2206445
5. Hamzaoglu F, Stan M (2001) Split-path skewed (SPS) CMOS buffer for high performance and low power applications. IEEE Trans Circuits Syst II 48(10):998–1002. https://doi.org/10.1109/82.974792
6. Jansson C, Chen K, Svensson C (1994) Linear, polynomial and exponential ramp generators with automatic slope adjustment. IEEE Trans Circuits Syst I 41(2):181–185. https://doi.org/10.1109/81.269058
7. Ke X, He L, Ma DB (2016) Design of high accuracy high conversion ratio single-stage switching power converter for modern FPGAs at 10MHz. In: 2016 13th IEEE international conference on solid-state and integrated circuit technology (ICSICT), pp 73–76. https://doi.org/10.1109/ICSICT.2016.7998842
8. Li N, Haviland G, Tuszynski A (1990) CMOS tapered buffer. IEEE J Solid-State Circuits 25(4):1005–1008. https://doi.org/10.1109/4.58293
9. Liu Z, Lee H (2013) A 100V gate driver with sub-nanosecond-delay capacitive-coupled level shifting and dynamic timing control for ZVS-based synchronous power converters. In: Proceedings of the IEEE 2013 custom integrated circuits conference, pp 1–4. https://doi.org/10.1109/CICC.2013.6658482
10. Liu Z, Cong L, Lee H (2015) Design of on-chip gate drivers with power-efficient high-speed level shifting and dynamic timing control for high-voltage synchronous switching power converters. IEEE J Solid-State Circuits 50(6):1463–1477. https://doi.org/10.1109/JSSC.2015.2422075
11. Liu D, Hollis SJ, Dymond HCP, McNeill N, Stark BH (2016) Design of 370-ps delay floating-voltage level shifters with 30-V/ns power supply slew tolerance. IEEE Trans Circuits Syst II 63(7):688–692. https://doi.org/10.1109/TCSII.2016.2530902

12. Ma H, van der Zee R, Nauta B (2013) An integrated 80-V class-D power output stage with 94% efficiency in a 0.14 μm SOI BCD process. In: 2013 Proceedings of the ESSCIRC (ESSCIRC), pp 89–92. https://doi.org/10.1109/ESSCIRC.2013.6649079
13. Maderbacher G, Jackum T, Pribyl W, Michaelis S, Michaelis D, Sandner C (2011) Fast and robust level shifters in 65 nm CMOS. In: 2011 Proceedings of the ESSCIRC (ESSCIRC), pp 195–198. https://doi.org/10.1109/ESSCIRC.2011.6044898
14. Moghe Y, Lehmann T, Piessens T (2011) Nanosecond delay floating high voltage level shifters in a 0.35 μm HV-CMOS technology. IEEE J Solid-State Circuits 46(2):485–497. https://doi.org/10.1109/JSSC.2010.2091322
15. Takai N, Fujimura Y (2007) Compensation method of amplitude error in sawtooth wave generator. In: 2007 18th European conference on circuit theory and design, pp 128–131. https://doi.org/10.1109/ECCTD.2007.4529553
16. Takai N, Fujimura Y (2008) Sawtooth generator using two triangular waves. In: 2008 51st Midwest symposium on circuits and systems, pp 706–709. https://doi.org/10.1109/MWSCAS.2008.4616897
17. Wittmann J, Wicht B (2013) MHz-converter design for high conversion ratio. In: 2013 25th International symposium on power semiconductor devices & IC's (ISPSD), pp 127–130. https://doi.org/10.1109/ISPSD.2013.6694445
18. Wittmann J, Rosahl T, Wicht B (2014) A 50V high-speed level shifter with high dV/dt immunity for multi-MHz DCDC converters. In: European solid state circuits conference (ESSCIRC), ESSCIRC 2014 – 40th, pp 151–154. https://doi.org/10.1109/ESSCIRC.2014.6942044

Chapter 5
Efficiency and Loss Modeling of High-V_{in} Multi-MHz Converters

In this chapter, an efficiency model suitable especially for multi-MHz high-V_{in} converters is proposed. It allows to optimize design parameters, like on-state resistance, gate supply voltage, switching frequency, etc., and to study and compare various converter architectures in terms of efficiency. For the analysis of the converter, it is essential to determine both the loss causing elements (loss causes) and the elements, in which the losses are finally dissipated (loss locations).

Section 5.1 describes conventional efficiency models and their limitations with respect to the purpose of this book. Section 5.2 shows the priorities of the implementation of the proposed efficiency model. Valid approximations of the loss contributors are studied for an asynchronous buck converter in Sect. 5.3 and a synchronous buck converter in Sect. 5.4. While in an asynchronous buck converter, the main switching losses are caused by the switching transitions and the parasitic capacitances of the switch and the Schottky diode, in a synchronous converter the additionally dominating dead time related losses are discussed.

In Sect. 5.6, an efficiency comparison is done between an asynchronous and a synchronous buck converter to demonstrate the influence of the dead time related losses on the overall buck converter architecture selection.

Section 5.7 introduces a design indicator, with the purpose to predict the efficiency of a converter architecture if it would be operated at different switching frequency, input voltage, output voltage, or output current. The design indicator allows thus a first-order comparison of state-of-the-art converters, which are typically shown at different operating points.

© Springer Nature Switzerland AG 2020
J. Wittmann, *Integrated High-V_{in} Multi-MHz Converters*,
https://doi.org/10.1007/978-3-030-25257-1_5

5.1 Conventional Efficiency Modeling

Several efficiency models are published for asynchronous and synchronous buck converters [4–12, 14–17]. An overview of the covered losses in each of the available references is shown in [9]. Most of the published models are suitable for low switching frequencies, high input voltages, and larger currents with external power switches. Several efficiency models determine the transition time, and thus the related transition losses only by covering the current through C_{gd} of the power switch (miller capacitance) during the voltage transition at the drain, while the contribution of the voltage transition at the gate is neglected. Moreover, the non-linear voltage dependence of the capacitances is ignored, or roughly linearized [5]. Only [12] includes a detailed analytical model for the voltage dependent capacitances. References [8] and [12] show that parasitic inductances contribute significantly to losses at high output current (>10 A) and external power switches. Loss models for synchronous buck converters typically cover body diode conduction losses during dead time. Reverse recovery losses are only added in [6] and [11] as a fixed reverse recovery charge. However, if the dead time becomes as small as a few nanoseconds, reverse recovery charge Q_{rr} starts to depend also on the duration of the body diode conduction, and thus on the dead time, as well as on the current through the body diode, and thus the converter's output current. Furthermore, none of the reference includes the losses caused by the generation of the high-side gate driver supply, which can be a large contributor at high input voltages and fast-switching gate drivers.

As the state-of-the-art models are mostly published for converters with low switching frequency and external power switches, the required accuracy of each loss contributors needs to be re-evaluated and improved for fast-switching high-V_{in} converters.

5.2 Priorities of Loss Modeling for High-V_{in} Fast-Switching Converters

To predict and analyze the efficiency of the converters covered in this book, an efficiency model was implemented, which is suitable to cover precisely the losses in multi-MHz high-V_{in} converters. In the following, the main loss contributors are analyzed and approximations for the efficiency models are discussed.

In multi-MHz converters, switching losses become dominant and the transition time has to be significantly faster (see Sect. 2.3.5 and Fig. 2.18). The non-linearity of the parasitic capacitances significantly influences the losses. Dead time related losses have to be covered more accurately as they increase with both V_{in} and the switching frequency. In addition, a split up of the elements which cause the losses (e.g., capacitances) and the loss locations at which the losses are finally dissipated (e.g., the resistance of the power switches) is desired for analysis purpose.

The constraint of having an integrated power switch, with relatively low output current of up to 1–2 A, leads to a shift of the focus of the loss elements to be modeled. The efficiency study of this book results in the following priorities for the model implementation, which are valid for both, asynchronous and synchronous buck converters.

Switch conduction losses are calculated as typically done in conventional models (see also Sect. 2.3.5), while the on- and off-times are determined based on the ideal duty cycle $D = t_{on}/t_{off} = V_{out}/V_{in}$. At very high switching frequencies, the influence of the finite switching node transitions becomes dominant, which results in a slightly different effective duty cycle at the switching node, and a slight error in the on- and off-time calculation is expected. As fast-switching converters with a high conversion ratio have to be designed for very short switching transitions to achieve a minimum required on-time pulse at the switching node (compare Sect. 2.3.6, Fig. 2.18), the influence on the conduction loss calculation remains small. The falling transition time becomes large, as soon as the switching node is slowly discharged by a small inductor current I_{L0} at light loads. This is the case if a low-side diode is used (asynchronous buck), or the low-side switch (synchronous buck) remains off until the switching node is discharged to ground. The error in the duty cycle thus is larger at smaller loads, however, the influence of the conduction losses at small loads is again negligible and justifies to use simply the ideal duty cycle for conduction loss calculations. At high input voltages, the conduction losses in fast-switching converters are typically less dominant, compared to the losses caused by parasitic capacitances, as there is a design limitation of a maximum possible on-state resistance. The design of the on-state resistance of the switches is discussed in more detail in Sect. 5.5.

Transition losses have a very specific behavior with the strong gate driver and the low gate resistances of the integrated switches used in this book. The Miller plateau becomes very short, and the non-linearity of the drain capacitance dominates the transition behavior and the transition time. The transition of the switching node and its impact on the losses are discussed in detail in Sect. 5.3.

Losses caused by parasitic capacitances of the switches, especially at the drain, are very critical as they increase with the input voltage and the switching frequency. As shown in Table 3.1 (see Sect. 3.2), the parasitic capacitances at a given on-state resistance are up to 5–10 times higher in fully integrated switches. A more accurate modeling of the parasitic capacitances and their non-linearity is necessary. The high non-linearity of the capacitances, which are different for example in triode and saturation region of the switch, have a crucial influence on the calculated efficiency. A more detailed analysis of the parasitic capacitances is shown in Sect. 5.3.

Inductor losses are not covered in the efficiency model as they can be calculated independently of the power stage. The inductor losses are used from the vendors model, while the inductors used for the converters in this book are selected based on the inductor study of Sect. 2.3.3.

Losses at the equivalent series resistance of the output capacitor R_{esr} are neglected, as a voltage mode Type III compensation is used for the buck converters in this book (see Sect. 4.1), which allows to use output capacitors with a small R_{esr} [1].

Parasitic inductances are neglected as integrated power switches are used, and the current levels are low. For example, a voltage drop over a bond wire inductance of <1 nH at a switched current slope of 0.5 A/ns would result in a voltage drop of <1 V over the bond wires, assuming an optimized chip connection to the PCB (for example direct-bond to PCB, as shown in Fig. 3.6).

Gate driver and control circuit losses are extracted and scaled with the switching frequency. The gate driver losses do not include the losses of the last driver stage, as they are included in the charging losses of the gate capacitance of the power switches.

Losses of the gate supply generation are caused in the circuits which provide the supply voltage for the gate driver at the low side, and the supply voltage for level shifter and gate driver at the high side. The low-side gate driver is typically supplied by a separate regulator for a low-voltage supply rail, which is anyway available to supply low-voltage components in the system. The efficiency η_{ls} of this converter can be considered by multiplying the gate driver losses by a factor of $1 + \eta_{ls}$. Sometimes, the output voltage of the buck converter itself can be used (after start-up completed) if its output voltage fits the required gate supply voltage. For the efficiency studies in this book, the low-voltage gate driver supply is assumed to be generated ideally for better comparison purpose, as η_{ls} from the system's low-voltage converter is unknown. The high-side supply for the level shifter and the gate driver has to be generated separately as the supply rail has to follow the input voltage or the switching node, for PMOS or NMOS switches, respectively. An integrated linear regulator, which is often used to supply the gate driver of PMOS switches, as shown in Fig. 3.3a, has to be fully considered, as the losses are significantly due to the high voltage drop over the pass device (MP_{lin}). In case, a charge pump (Fig. 3.3b) or a bootstrap circuit (Fig. 3.3c) is used, the charge pump or bootstrap efficiency has to be considered. The input voltage of the charge pump or bootstrap circuit (V_{dd5}) can be again the low-voltage system supply, and is thus to be assumed as ideal in the efficiency comparisons in this book.

5.3 Efficiency Model for an Asynchronous Buck Converter

In the following, the main aspects of the implementation of the efficiency model are revealed, based on an implementation of an asynchronous buck converter, achieving switching frequencies beyond 10 MHz and input voltages up to 50 V. This contains the impact of the freewheeling diode, the non-linear parasitic capacitances, and especially their influence on the transition losses.

5.3.1 Diode Conduction Losses

As a low-side diode, a SS16 Schottky diode [13] was chosen for the implementation of the asynchronous buck converter in this book to avoid reverse recovery losses. The SS16 is a typically used diode in automotive applications, which can withstand blocking voltages up to 60 V, and average forward currents up to 1 A. The advantage of the asynchronous buck converter is that no low-side gate driver is required, and thus switching losses are lower. As a drawback, higher losses $V_f \cdot I_{L0}$ occur in the Schottky diode, which is the voltage drop caused by the inductor current over the forward voltage V_f of the diode. V_f is typically higher, than the voltage drop V_{ds} in a low-side switch with a low on-state resistance. In the efficiency model, V_f is assumed as constant value, which is extracted at a medium diode current (for example, $V_f = 0.415$ V at a diode current of 250 mA). V_f slightly increases for higher forward currents. However, using a constant V_f for all currents introduces only a negligible error, as for higher currents a self-heating of the diode occurs, as depicted in [13]. Higher diode temperatures reduce V_f and thus compensate an increasing V_f due to higher currents, which justifies the use of a constant V_f as a first-order approximation.

5.3.2 Switching Behavior

Figure 5.1 shows a simulation of the switching transition of an asynchronous buck converter for the rising and the falling switching transition, at a converter load current of $I_{out} = 0.5$ A. Unless conventional buck converters with moderate switching frequencies, the buck converter in this book is designed to operate at switching frequencies above 10 MHz and high conversion ratios up to $V_{in}/V_{out} = 10$ and larger. The required minimum on-time pulse of the high-side switch in the range of 3 ns, as derived in Sect. 2.3.6 (Fig. 2.18) is only achieved with a very fast gate transition. The prototype for this book was designed to have a gate transition in the range of 1 ns (see gate driver design in Sect. 4.4), which can be observed at the gate-source voltage V_{gs} of the high-side switch at turn-on in phase 1, as depicted in Fig. 5.1a (top). The bottom part of Fig. 5.1a shows the drain current I_d of the high-side switch and the diode forward current I_{dio}. Instantly after V_{gs} of the high-side switch reaches the threshold voltage, the current commutates from the diode to the high-side switch within approximately $t_{ri} \approx 100$ ps. t_{ri} is thus short enough for the related current transition losses to be neglected, and the current to be assumed to commutate instantly. After the current transition t_{ri}, the current I_d in the high-side switch rises up to approximately 3.5 A, at which the high-side switch saturates. In this period, I_d delivers the inductor current of $I_{L0} \approx 0.5$ A. The remaining current (≈ 3 A) is flowing into the diode (Fig. 5.1a, bottom) and thus charging its parasitic capacitance. However, even with a high current flowing into the diode, the switching node voltage V_{sw} remains low during the time interval t_{vi}.

Fig. 5.1 Simulation of
(**a**) the rising transition and
(**b**) the falling transition of a
multi-MHz high-V_{in} buck
converter

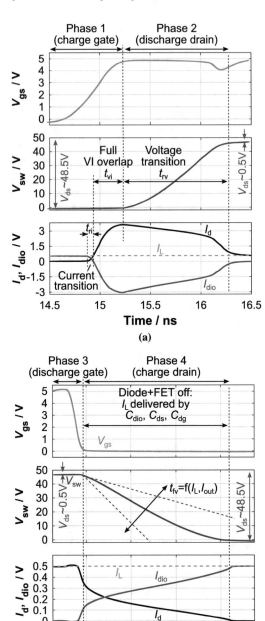

Fig. 5.2 Parasitic junction capacitance of the SS16 Schottky diode in dependence of the reverse voltage

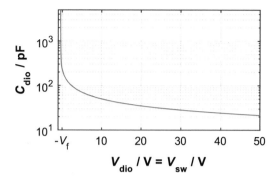

The unexpected behavior of the switching node staying below zero during the interval t_{vi} leads to excessive losses, as the inductor current, or respectively the output current, suffers from a voltage drop over the switch of the full input voltage V_{in}.

This effect of V_{sw} staying low during t_{vi} can be explained with the non-linearity of the parasitic capacitance of the Schottky diode. The capacitance of the Schottky diode SS16 is plotted versus the reverse blocking voltage V_{dio} (which is V_{sw} in the buck converter) in Fig. 5.2.

The capacitance of the Schottky diode is formed by the depletion region, which strongly depends on V_{dio} and thus V_{sw}. C_{dio} is low in a range of 20–70 pF if V_{sw} is high. However, C_{dio} significantly increases by nearly three decades to the nanofarad range if V_{sw} decreases towards the forward bias voltage $-V_f$ of the diode.

As a consequence for the rising transition, a large amount of charge is required to bring the switching node from the forward conduction voltage $-V_f$ towards positive voltages during the rising transition (Fig. 5.1a), and V_{sw} appears to stay low during t_{vi}.

The time t_{vi} results from the finally available charging current of C_{dio}, and thus on the saturation current of the high-side switch, as well as on the inductor current (which can be assumed to be the load current for low or moderate inductor current ripples).

Simulations showed that the time t_{vi} (Fig. 5.1a) mainly depends on the inductor current I_{L0}. Higher load current leads to a proportionally higher inductor current I_{L0}, which is subtracted from the current delivered from the high-side switch during t_{vi}. Thus, the remaining current charging C_{dio} decreases. The diode is charged slower and t_{vi} increases.

Only after t_{vi}, V_{sw} starts to rise towards V_{in} during t_{rv} in a manner, which can be approximated linearly in the efficiency model. t_{rv} is directly proportional to the input voltage (final voltage of C_{dio}), which mainly determines t_{rv}. The influence of the varying load/inductor current is small, as the variation of the load current is only in the range of 10–20% of the available saturation current from the high-side switch, charging the switching node capacitances. As the losses become less dominant if V_{sw} rises towards V_{in}, the influence of I_{out} on t_{rv} can be neglected in the model.

This analysis leads to a proposal to linearize the overall rising transition time in the efficiency model as

$$t_r = \underbrace{t_{vi0,a} + t_{vi0,b} \cdot \frac{I_{out}}{I_{out,tr0}}}_{t_{vi}} + \underbrace{\frac{1}{2} \cdot t_{rv0} \cdot \frac{V_{in}}{V_{in,tr0}}}_{t_{rv}} \tag{5.1}$$

The parameters $t_{vi0,a}$, $t_{vi0,b}$, $I_{out,tr0}$, t_{rv0}, and $V_{in,tr0}$ are fitted to the results of a parametric simulation of the transition time, varying I_{out} and V_{in}.

With the linearized transition time t_r, the rising transition losses P_{tr} are calculated in a conventional way to

$$P_{tr} = f_{sw} \cdot V_{in} \cdot I_{out} \cdot t_r. \tag{5.2}$$

The falling transition of the asynchronous buck converter is shown in Fig. 5.1b. The gate transition of V_{gs} occurs again within 1 ns, and the switch is assumed to close instantly. In the following period (t_{rv}), in which V_{sw} is discharge towards ground, the inductor current is supplied by the parasitic capacitances at the drain of the high-side switch (I_d) and the Schottky diode (I_{dio}). The discharge time thus depends on the input voltage and the inductor current and thus the load current. For the asynchronous converter, no losses occur in the output stage during t_{fv}, as the charge of the parasitic capacitances at the switching node is discharge ideally lossless by the inductor, which delivers the parasitic charge to the output as load current.

5.3.3 Implementation as Four-Phase Model

As observable in Fig. 5.1a, the voltage transition at the gate is completed until V_{sw} noticeable starts to change in both the rising and falling transitions. This justifies to calculate the capacitive losses separately for the gate and the V_{sw} transitions. This results in four phases, two for each the rising and the falling transition. Each transition causes one or several capacitances to be charged or discharged. Each phase allows to analyze the involved parasitic capacitances to be charged or discharged. The amount of charge, circulating in the power stage during the according transition is determined by the size of the capacitances charged by the amplitude of the transition. The current paths in the power stage and the involved capacitors are depicted in Fig. 5.3, separately for each phase, and allow to depict the involved capacitors in each phase:

In phase 1, the high-side switch turns on, while V_{sw} is constant at $V_{sw} \approx 0$ ($V_{ds} \approx V_{in}$). The gate of the switch is charged, and thus C_{gd} (gate-to-drain capacitance) is discharged by the gate driver from $V_{sw} - V_d$ ($\approx 0 - V_{in}$) to $V_{boot} - V_d$ ($\approx 5\,V - V_{in}$), while C_{gs} (gate-to-source capacitance) is charged from V_{sw} (≈ 0) to the gate driver supply V_{boot} ($\approx 5\,V$).

Fig. 5.3 Charging and discharging current paths of the asynchronous buck converter with NMOS high-side switch in its four switching phases (phase 1–4, (**a**)–(**d**))

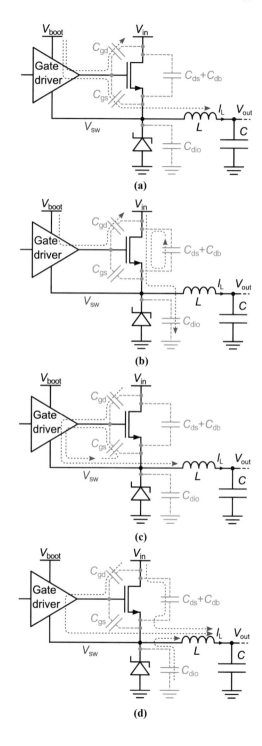

In phase 2, the gate of the switch is tight to V_{boot}, and the voltage of the switching node V_{sw} transitions from GND to V_{in}; V_{ds} is discharged from $V_{in} + V_f$ to the forward voltage drop $R_{on} \cdot I_{L0}$. C_{gd}, C_{ds} (drain-to-source capacitance), and C_{db} (drain-to-bulk capacitance) are discharged by the high-side switch, and C_{dio} is charged from $-V_f$ to V_{in}.

In phase 3, the gate of the high-side switch is discharged, while V_{sw} is constant close to V_{in} ($V_{ds} = R_{on} \cdot I_{L0}$). C_{gd} and C_{gs} are discharged from V_{boot} to V_{sw}.

In phase 4, the gate of the high-side switch is low, while V_{sw} is discharged by the inductor current to $-V_f$; V_{ds} transitions from $R_{on} \cdot I_{L0}$ to $V_{in} + V_f$.

By charging or discharging the capacitances, losses occur. However, the losses are not dissipated in the capacitances, but in the resistive elements in the charging or discharging current path (for example, gate driver or power switch). Anyway, the amount of losses are determined by the capacitances, which are generating the charge in the current paths.

In phase 1, for example, the amount of charge flowing to C_{gs} can then be calculated as $Q_{gs,1} = C_{gs} \cdot (V_{boot} - V_{sw})$, while $V_{boot} - V_{sw}$ is the gate driver supply. The same is done for C_{gd}, resulting in $Q_{gd,1} = C_{gd} \cdot (V_{boot} - V_{sw})$. The losses resulting from charging the gate with $Q_{gs,1} + Q_{gd,1}$ are dissipated in the resistance of the pull-up transistor of the gate driver output. In phase 2, as a second example, most of the losses are dissipated in the high-side switch while discharging C_{ds} and charging C_{dio}. However, C_{gd} is discharged through the high-side transistor of the gate driver output, as V_{boot} follows the rising V_{sw}-transition. Gate driver losses occur even without a transition of V_{gs}. In phase 4, the discharge of the capacitances is caused by the inductor current, and thus ideally no losses occur as the energy is transferred from the capacitances to the inductor, and finally to the output with ideally no losses.

5.3.4   Root Cause and Loss Location Analysis

The implementation of the phases with the analysis of the current paths in the efficiency model allows both, to reference the capacitive losses to the capacitances to distinguish the root cause of the losses, and alternatively to the resistive elements where losses are dissipated, namely the loss locations. The dissipated losses in the resistive elements are caused by the charge in the charging current paths, generating a voltage drop over the resistive elements. Figure 5.4 shows a loss break down for the buck converter, generated with the implemented loss model. In Fig. 5.4a, the losses are split up by the root causes, while in (Fig. 5.4b), the losses are assigned to its locations, both with varying input voltages up to 50 V at a switching frequency of 10 MHz. This allows a detailed analysis, for example, to determine the accurate heat dissipation in the power switch, and to observe that the power dissipated in the switch is mainly caused by the capacitance of the Schottky diode in this design.

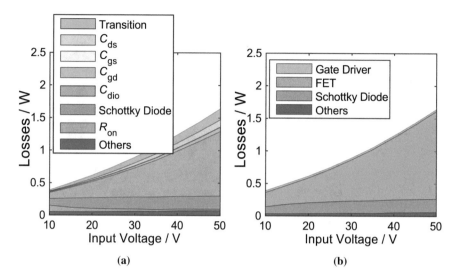

Fig. 5.4 Loss analysis by separation of (**a**) loss causes and (**b**) loss locations for an asynchronous buck converter with NMOS high-side switch at $f_{sw} = 10\,\text{MHz}$, $I_{out} = 0.5\,\text{A}$ and $V_{out} = 5\,\text{V}$

5.3.5 Contribution and Modeling of Non-linear Capacitances

For capacitors with constant capacitance, it is well known that during charging from zero to a capacitor voltage V_c, a charge $Q_c = C_c \cdot V_c$ is flowing to the capacitor. Assuming Q_c is delivered by the voltage source V_{src}, the energy W_{src} delivered by V_{src} during charging is

$$W_{src} = Q_c \cdot V_{src} = C_c \cdot V_c \cdot V_{src}. \tag{5.3}$$

The energy stored on a capacitor (W_c) can be determined with the help of the charge–voltage (QV) diagram, which is plotted in Fig. 5.5. For a constant C_c, the charge on the capacitor Q_c increases proportional (linear) with the capacitor voltage V_c. Charging a capacitor means that some charge ΔQ_c is added to the existing charge Q_c at the actual capacitor voltage $V_c(Q_c)$. V_c depends on the actual amount of charge Q_c on C_c. The energy ΔW_c, which is added to C_c can be thus calculated by $\Delta W_c = \Delta Q_c \cdot V_c(Q_c)$. Thus, the total energy W_c on C_c is obtained by summing all infinitesimal small charge steps dQ_c which were added to C_c at their according capacitor voltage up to the final total charge Q_1 on C_c, which results in the integral

$$W_c = \int_0^{Q_1} V_c(Q_c) dQ_c. \tag{5.4}$$

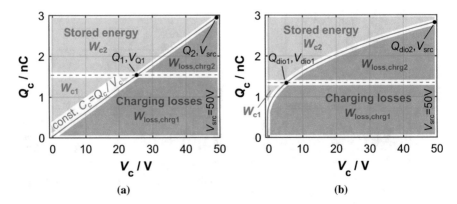

Fig. 5.5 Charge–voltage (QV) diagrams of (**a**) a constant capacitor, and (**b**) the Schottky diode SS16

Writing $V_c(Q_c)$ in relation of C_c and Q_c, and solving the integral, results in the well-known equation for the energy on constant (voltage independent) capacitors

$$W_c = \int_0^{Q_1} \frac{Q_c}{C_c} dQ_c = 0.5 \cdot \frac{Q_1^2}{C_c} = 0.5 \cdot C_c \cdot V_c^2. \qquad (5.5)$$

In case C_c is fully charged up to V_{src} ($V_c = V_{src}$), the energy from the supply (5.3) is $W_{src} = C_c \cdot V_{src}^2$. Equation (5.5) shows that for a fully charged capacitor ($V_c = V_{src}$), half of the energy is stored on the capacitor. As a result, the other half is dissipated as loss during charging. In general, the charging losses of a capacitor can thus be calculated as

$$W_{loss,chrg} = W_{src} - W_c. \qquad (5.6)$$

The stored energy W_c is dissipated as losses, when C_c is discharged again to zero. In the QV diagram in Fig. 5.5a, the stored energy calculated in (5.5) results in the area above the QV curve, while the charging losses calculated in (5.6) result in the area below the QV curve.

Figure 5.5a shows the loss and stored energy for two cases if the capacitor is charged from $V_{src} = 50\,V$ to $V_{Q1} = 0.5 \cdot V_{src}$ corresponding to stored charge Q_1, and if the capacitor is fully charged to V_{src}, resulting in a stored charge Q_2. During charging up to V_{Q1}, charging losses $W_{loss,chrg1}$ are much higher than the stored energy W_{c1}, as the charge is delivered from a considerably larger voltage V_{src} (compare (5.3)). In contrast, when charging from V_{Q1} to V_{src}, the charging losses $W_{loss,chrg2}$ are minor due to the decreasing voltage drop of the delivered charge, and the stored energy is significantly higher. This demonstrates that the majority of charging losses occur already until a capacitor is charged up to only half of it's voltage, and the majority of energy, which is finally stored on the capacitor, is transferred to the capacitor only at higher capacitor voltages.

As shown in Fig. 5.2, a real diode blocking capacitance, as well the parasitic capacitances of a switch, are depending on its current voltage and thus are highly non-linear. Figure 5.5b shows the QV diagram (extracted from the diode model) of the SS16 diode. The non-linearity of C_{dio} of the Schottky diode requires a large amount of charge (approximately 1nC) until the voltage starts to noticeably increase from $-V_f$ towards positive voltages. At higher V_{dio}, a lower capacitance requires only a small increase in the charge to increase V_{dio}. This means, in the energy calculation, the constant capacitance C_c in (5.5) becomes a voltage dependent capacitance $C_{dio}(V_{dio})$, and the QV curve changes to a non-linear curve as depicted in Fig. 5.5b. The strong increase of charge at low V_{dio} unbalances the charging losses $W_{loss,chrg}$ and the stored energy W_c. The charging loss area increases, and the stored energy decreases. When the diode is charged from $-V_f$ to only 5 V (Q_{dio1}, V_{dio1}), the charging losses $W_{loss,chrg1}$ are twice as large as the charging losses $W_{loss,chrg2}$, which occur during the remaining charging from V_{dio1} up to $V_{src} = 50$ V (Q_{dio2}). In contrast, the stored energy on the diode capacitance at $V_{dio1} = 5$ V is negligible; the main energy stored on C_{dio} is only accumulated at higher voltages of V_{dio}.

This analysis has a crucial meaning if soft-switching techniques are applied as describe in Sect. 2.2.2.4. Even a poor soft-switching condition, in which the switching node V_{sw} is pulled up by an ideal current source I_{pu} (as described in Sect. 2.2.2.4) only by a few volts before the high-side switch turns on, the major charging losses are already eliminated, and an ideal zero-voltage switching is not necessarily required.

A linear (constant) equivalent capacitor of the diode is typically obtained by using the final charge and the final voltage (Q_{dio2} and V_{src} in Fig. 5.5b), and calculating a fixed, voltage independent capacitor from this values. The QV diagram of the calculated equivalent capacitor would be identical to the behavior of the constant capacitor in Fig. 5.5a. By comparing the charging loss area of the equivalent capacitor of Fig. 5.5a to the real behavior in Fig. 5.5b, it becomes obvious that a large loss calculation error is induced.

Due to the high impact on the efficiency and loss accuracy, also the non-linearity of the capacitances of the power switch has to be considered. Figure 5.6 shows the non-linearity of the parasitic capacitances of the power switch in the particular switching phases 1–4 (Fig. 5.3a–d). The parasitic capacitances change, whether the switch is in off- or in on-state, and thus the capacitances in phase 1 and 2 are different to the capacitances in phase 3 and 4.

The gate capacitances in phase 1 shown in Fig. 5.6a (switch turns on and gate is charged, while $V_{ds} \approx V_{in}$) show a quite constant behavior across the gate driver supply range (typical 0–5 V), while in phase 3 (switch turns off, the gate discharges, while $V_{ds} = I_{L0} \cdot R_{on} \approx 0$ V), the capacitance vary by a factor of two. The gate capacitances are rather constant, compared to the depletion capacitances of the diode and at the drain. Moreover, due to the small charging amplitude, compared to V_{in} at the drain, the loss contribution of the gate charge is not significant, as it is confirmed by the loss break down of Fig. 5.4. Thus, C_{gs} and C_{gd} have been chosen to be each modeled as a constant capacitance, which is the averaged capacitance in each phase. The averaging is done separately for phase 1 and 3, as the capacitance values change.

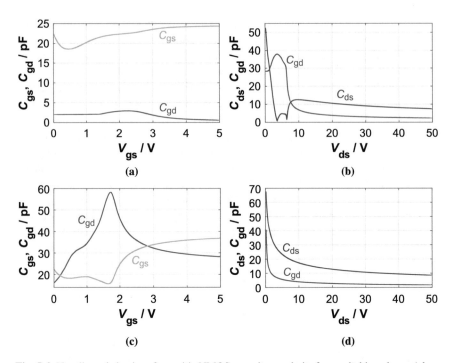

Fig. 5.6 Non-linear behavior of parasitic NMOS capacitances in its four switching phases (phase 1–4, (**a**) Phase 1: Transition of V_{gs} from 0 to 5 V, $V_{ds} = 0$ V. (**b**) Phase 2: Transition of V_{ds} from V_{in} to 0 V, $V_{gs} = 5$ V. (**c**) Phase 3: Transition of V_{gs} from 5 to 0 V, $V_{ds} = V_{in}$. (**d**) Phase 4: Transition of V_{ds} from 0 V to V_{in}, $V_{gs} = 5$ V)

The average of C_{gd} and C_{gd} is related to the total charge flowing from the gate driver supply (V_{boot}) to the gate, and back to the gate driver ground, which is V_{sw}. As the losses at the gate transition are finally defined by the total charge flowing from V_{boot} to V_{sw}, the averaging does not induce a calculation error of the total gate charge losses within one period. However, a separation of the modeled losses, which occur during charging and discharging the gate, respectively, might be inaccurate due to the non-linearity of C_{gd} and C_{gs}, as they slightly unbalance the charging losses and stored energy (discharging losses) caused by the gate capacitances, like in the Schottky diode as discussed in Fig. 5.5b.

Figure 5.6b shows the transition of V_{sw} in phase 2 (V_{gs} is constant at 5 V). During the V_{ds} transition from V_{in} down close to $V_{ds} = 5$ V, the switch is in saturation region; it can be observed that C_{gd} and C_{ds} are only slightly dependent on the voltage (comparing to the strong dependence of the Schottky diode in Fig. 5.2). Below $V_{ds} = 5$ V, the switch enters the linear region, which implies a strong change in the switch capacitances. The relatively constant capacitances allow to model C_{gd} and C_{ds} as averaged constant capacitors in the saturation region. Also in the linear region, constant capacitors averaged from 0–5 V are used in the model. In linear region, the capacitances are quite non-linear, a loss calculation error is expected. However, this error is negligible, as the discharge happens at a very low voltage V_{ds}, and thus the loss contribution of the linear region to the overall losses is small.

The behavior of the capacitances in phase 4 is depicted in Fig. 5.6d, while the switch is turned off ($V_{gs} = 0$ V). A stronger voltage dependence, especially of C_{ds}, can be observed at the drain capacitance, which are more closely to the behavior of the Schottky diode. This would require also a more accurate modeling by integrating the charge as shown in (5.5). However, in an asynchronous converter, the switching node is discharged (and thus V_{ds} is charged) by the inductor current. No charging losses occur, as the energy during charging (compare to Fig. 5.5b) is not dissipated in a resistive element, but is transferred into the inductor energy, where it is finally used as output current.

The accuracy of the efficiency model for an implemented 10 MHz-switching asynchronous buck converter is shown in Fig. 5.7. Figure 5.7a shows the result of the efficiency model, in which C_{dio} is accurately modeled by integrating the energy along the QV curve. In Fig. 5.7b, a simplified model uses only a constant capacitor for C_{dio}, which is calculated as $C_{dio} = Q_{dio}(V_{in})/V_{in}$. A comparison to the results to the efficiency measurements of Fig. 5.7c confirms that modeling the non-linear capacitances as voltage independent constant capacitors only, results in a large error in the efficiency calculation, while the non-linear model matches the measured results very well. The efficiency model with the proposed priorities for modeling the losses of high-V_{in} multi-MHz converters matches the measured efficiency of the buck converter with a below 3% accuracy over a wide input voltage range up to $V_{in} = 50$ V.

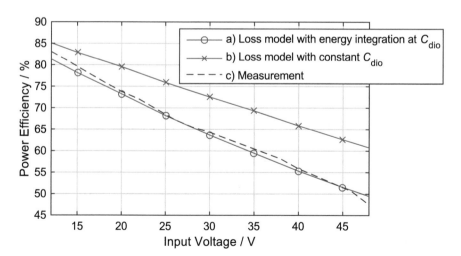

Fig. 5.7 Efficiency comparison of the proposed efficiency model for an asynchronous buck converter with $f_{sw} = 10$ MHz, $V_{out} = 5$ V and $I_{out} = 300$ mA, comparing (a) the model with accurate C_{dio} obtained by energy integration, (b) the model with constant C_{dio} leading to large efficiency error, and (c) efficiency measurements of the implemented buck converter

5.4   Loss Contributions in a Synchronous Buck Converter

As introduced in Sects. 2.3.5 and 3.1, the main differences of the synchronous compared to the asynchronous buck converter is the use of a low-side switch, instead of a freewheeling diode. The diode loss model has to be replaced by the loss model of the low-side switch, and an additional low-side gate driver is contributing to the losses. Moreover, the required dead time between the high-side and low-side switch conduction adds significant losses, which are subsequently described.

5.4.1   Contribution of the Low-Side Switch

The benefit of using a low-side switch is that it can be designed to have lower conduction losses, as the on-state resistance of the switch can be designed to have an arbitrary small voltage drop, while the minimum voltage drop of a diode is always limited to its forward conduction voltage V_{f}. As a drawback, the gate driver for the low-side switch adds additional losses, while a lower on-state resistance with a larger switch results in higher gate driver losses. A trade-off has to be found between gate driver losses and on-state losses. To optimize conduction losses, the ratio of the on-state resistance of the high-side switch to the low-side switch can be sized according to the duty cycle of the buck converter, in case the conversion ratio of the converter is nearly constant. This is, because the conduction time of each switch is proportional to its losses, assuming the inductor current do not vary significant during the high-side and low-side on-state. However, the prototype of the buck converter proposed in this book shall be suitable to cover especially a widely varying input voltage. Thus, the size of the high-side and low-side switch, as well as their gate drivers, are implemented with the same size.

The parasitic capacitances of the switch in phase 1–3 (see Fig. 5.6) are modeled identically as for the high-side switch. The drain capacitances in phase 4 shown in Fig. 5.6d become dominant in loss contribution, as C_{ds} and C_{gd} of the low-side switch in off-state are charged from $-V_{\text{f}}$ (or $R_{\text{on}} \cdot I_{\text{L0}}$, depending on the dead time as described in the subsequent section) to close to V_{in} when the high-side switch turns on. An accurate modeling of both capacitances with integrating the voltage dependent QV curve is required, as previously proposed for the Schottky diode in (5.5) and Fig. 5.5b.

When the high-side switch turns off, it strongly depends on the dead time, whether the switching node is discharged by the low-side switch or by the inductor current (lossless discharge). In case the low-side switch is turned on while the switching node is still high, it discharges the switching node to zero, while C_{ds} and C_{gd} of the high-side switch in off-state are charged (phase 4, Fig. 5.6d). Also in this case, the losses caused by C_{ds} and C_{gd} of the high-side switch require to be modeled accurately by charge integration.

5.4.2 Influence of Dead Time on Switching Behavior

The principle of the dead time and the impact on the switching behavior in a synchronous buck converter is depicted in Fig. 5.8 [18, 19]. Two dead times have to be considered, the dead time DT_{hi} between the low-side switch turn-off and the high-side switch turn-on (turn-high or rising transition of V_{sw}), as well as the dead time DT_{lo} between the high-side switch turn-off and low-side switch turn-on (turn-low or falling transition of V_{sw}).

To model the influence on the switching losses, two different cases have to be distinguished. Figure 5.8a shows the behavior of the switching node for a typical case, in which the load current is high enough that the inductor current, especially at its minimum current peak, remains positive. In Fig. 5.8b, the load current is assumed to be low enough, such that the inductor current becomes negative towards the end of the converter off-time, while the low-side switch is on. This is considered as light-load case in the following.

The fundamental difference of this two cases occurs after the low-side switch is turned off, i.e., during the dead time DT_{hi}. A positive inductor current, flowing out of the power stage pulls the switching node further down into body diode conduction of the low-side switch, while at light load, the negative current flows into the power stage, and pulls the switching node high towards V_{in}.

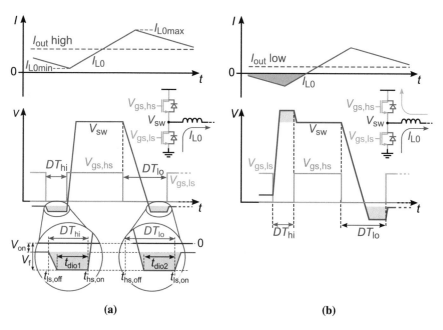

Fig. 5.8 Behavior of the switching node during dead time at (**a**) high output current, and (**b**) at low output current (light load)

At the falling transition of V_{sw} during DT_{lo}, the behavior of the switching node is identical, as in both cases (Fig. 5.8a, b) the inductor current is positive at the high-side switch turn-off.

5.4.3   Dead Time Related Losses

In the typical case, as shown in Fig. 5.8a, the dead time DT_{hi} starts at $t_{ls,off}$ (see enlargements in Fig. 5.8a), when the low-side switch turns off and the inductor current starts to pull V_{sw} further negative. The body diode of the low-side switch starts conducting, and V_{sw} is held at the negative forward voltage of the diode $-V_f$. At the end of DT_{hi} at $t_{hs,on}$, V_{sw} is pulled up towards V_{in} by the high-side switch.

The second dead time DT_{lo} has a similar scheme. At the beginning of DT_{lo}, at $t_{hs,off}$, when the high-side switch is turned off, V_{sw} is pulled down by the inductor current towards $-V_f$ until the diode conducts. At $t_{ls,on}$, at the end of DT_{lo}, the low-side switch is turned on.

5.4.3.1   Switching Loss Analysis

Figure 5.9 shows the switching behavior for three cases, both dead times can be either too large (Fig. 5.9a), at optimum (Fig. 5.9b), or too short (Fig. 5.9c). Is the dead time too large, the body diode of the low-side switch is forward biased due to the forced inductor current I_{L0} of L_0, and V_{sw} is pulled low to a negative diode forward voltage $-V_f$.

At the end of DT_{hi}, when the high-side switch turns on, the body diode of the low-side switch commutates from forward biased state to blocking state. A large

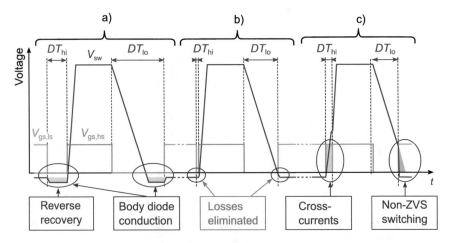

Fig. 5.9 Switching states of the dead time controller. (**a**) Dead time too large. (**b**) Optimal dead time. (**c**) Dead time too small

amount of reverse recovery charge Q_{rr} flows into the depletion region of the body diode to block the diode [2]. As Q_{rr} is delivered by the high-side switch from V_{in}, large losses occur. The optimal switching point (Fig. 5.9b) is achieved if the high-side or low-side switch, respectively, turns on a very short instant before the body diode of the low-side switch starts conducting. This eliminates forward conduction and, subsequently, reverse recovery losses. If the dead times are too small or even negative (Fig. 5.9c), cross conduction occurs, as both switches are turned on at the same time. This leads to excessive power loss and needs to be avoided in any case.

DT_{lo} should be large enough to avoid that the low-side switch turns on during the falling slope of V_{sw}. The losses are minimal if the low-side switch is turned on with zero-voltage across it, which is referred to as zero-voltage switching (ZVS), as introduced in Sect. 2.2.2.4. This can be achieved if the parasitic capacitances at the switching node C_{sw} are fully discharged by the inductor current while the low-side switch is still turned off. Thus, the charge of the parasitic capacitances contributes to the output current, and is not dissipated to ground by the low-side switch.

To determine the loss impact if DT_{hi} and DT_{lo} are not chosen to be at the optimum, and to determine the required accuracy of the dead time to avoid dead time related losses, the losses depending on DT_{hi} and DT_{lo} are analyzed separately in the following [16].

5.4.3.2 Losses Related to Turn-High Dead Time

At the turn on event during DT_{hi}, hard-switching with a large drain-source voltage at the high-side switch cannot be avoided in CCM [3, 20], but the reverse recovery and forward conducting diode losses can be significantly reduced, or even eliminated, with an optimal dead time. DT_{hi} should be large enough to avoid cross conduction in any case.

Figure 5.10a shows the dead time dependent losses versus varying dead time DT_{hi}, containing diode conduction losses and reverse recovery losses for input voltages of 12 and 18 V, at 5 V output and 10 MHz switching. After the low-side switch turns off at $t_{ls,off}$ (see Fig. 5.8a), the diode is not conducting immediately, as the switching node first needs to be further discharged to $-V_f$. This time is extracted from a transistor level simulation of the power stage, which is $DT_{hi} - t_{dio1} \approx 500$ ps in the operating point of Fig. 5.10a. During this time, no additional dead time related losses occur.

Towards larger dead times DT_{hi}, the losses increase due to larger diode conduction time t_{dio1} (see Fig. 5.8a). The losses shown in Fig. 5.10a are calculated by

$$P_{DThi} = t_{dio1} \cdot f_{sw} \cdot (\underbrace{I_{Lmin} \cdot V_f|_{@I_{Lmin}}}_{\text{Diode Conduction Losses}} - \underbrace{I_{Lmin}^2 \cdot R_{on,ls}}_{\text{LS-On-State Losses}})$$

$$+ \underbrace{Q_{rr}(t_{dio1}, I_{Lmin}) \cdot V_{in} \cdot f_{sw}}_{\text{Reverse Recovery Losses}}. \tag{5.7}$$

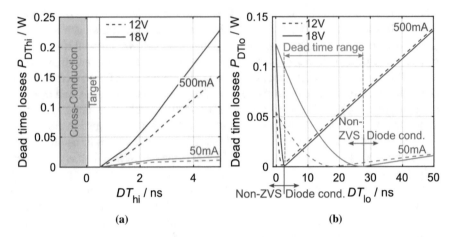

Fig. 5.10 Dead time dependent losses at different output currents I_{out} (50 and 500 mA) versus (**a**) turn-on dead time DT_{hi} and (**b**) turn-off dead time DT_{lo}

P_{DThi} covers the additional losses caused by the dead time. For optimized dead time, losses due to the on-state resistance $R_{on,ls}$ in the conducting low-side switch would occur instead of the losses in the conducting diode. As (5.7) considers dead time dependent losses only, the on-state losses have to be subtracted from P_{DThi}. Q_{rr} depends on the diode conduction time t_{dio1} and the current through the diode, which is I_{Lmin} in this case. The reverse recovery charge of the body diode Q_{rr} was extracted from a transistor level simulation with technology models for the reverse recovery effect.

The losses for varying dead times DT_{hi} are calculated for two different currents $I_{out} = 50$ mA and $I_{out} = 500$ mA. The results of the simulation for $I_{out} = 500$ mA show that an increase of the dead time of about 1 ns increases the losses by nearly ≥ 0.1 W. At a converter output voltage of $V_{out} = 5$ V, this corresponds to an efficiency decrease of 4%. Equation (5.7) shows that the reverse recovery losses scale up linearly with the input voltage V_{in}, and thus become dominant in high-V_{in} converters.

As a conclusion from Fig. 5.10a and (5.7), the dead time should be adjustable with a minimum dead time value of less than 500 ps to be in the target window. This is especially important for higher input voltages, as V_{in} scale up the dead time related losses.

5.4.3.3   Losses Related to Turn-Low Dead Time

The losses during the dead time DT_{lo} of the turn-off event are plotted in dependency of DT_{lo} in Fig. 5.10b. The ideal dead time is achieved if the low-side switch turns on with ZVS, and if no body diode conduction occurs, after the parasitic capacitances C_{sw} at the switching node are completely discharged by the inductor current I_{L0}. If DT_{lo} is too short, losses occur as the switching node V_{sw} is still positive at $t_{ls,on}$ (see Fig. 5.9c), when the low-side turns on. The discharge of the parasitic capacitance C_{sw} at the switching node by the inductor current I_{L0}, which can be approximated by I_{Lmax} during DT_{lo}, determines the voltage at V_{sw} at $t_{ls,on}$ as

$$V_{sw}(t_{ls,on}) = V_{in} - \frac{1}{C_{sw}} \cdot I_{Lmax} \cdot DT_{lo}. \tag{5.8}$$

Equation (5.8) is valid for short dead times, when the low-side switch turns on, before diode conduction occurs. During diode conduction, the switching node voltage is $V_{sw} = -V_f$. At turn-on of the low-side switch, V_{sw} is pulled to $-V_{on}$ (see Fig. 5.8a), which is the drain-source voltage of the conducting low-side switch after turn on. For minimum losses, the low side switch is turned on exactly, when $V_{sw} = V_{on}$. V_{on} is very small in practical designs and hence can be neglected for simplification. Therefore, switching at zero voltage is assumed to be optimal (ZVS).

In case DT_{lo} is too long, body diode conduction losses occur (see Fig. 5.9c). The total losses P_{DTlo} shown in Fig. 5.10b are calculated by

$$P_{DTlo} = \underbrace{\frac{1}{2} \cdot C_{sw} \cdot V_{sw}^2(t_{ls,on}) \cdot f_{sw}}_{\text{Non-ZVS Losses}} + \underbrace{I_{Lmax} \cdot t_{dio2} \cdot V_f \cdot f_{sw}}_{\text{Diode Conduction Losses}}$$

$$- \underbrace{I_{Lmax}^2 \cdot R_{on,ls} \cdot t_{dio2} \cdot f_{sw}}_{\text{LS-On-State Losses}}. \tag{5.9}$$

The on-state losses of the low-side switch are again subtracted, as for optimal dead time the low-side switch conducts instead of the body diode.

The results in Fig. 5.10b consider $I_{Lmax} = 50\,\text{mA}$ and $I_{Lmax} = 500\,\text{mA}$, as well as two different input voltages $V_{in} = 12\,\text{V}$ and $V_{in} = 18\,\text{V}$. A larger input voltage has a negligible effect on the diode conduction losses, but significantly increases the non-ZVS losses. For lower peak inductor current (lower I_{Lmax}), the optimal dead time point is shifted to a longer dead time, as the falling slope of V_{sw} decreases.

As a result, it is required to adjust the dead time over a wide range to eliminate dead time related losses if the operating point (V_{in} and I_{out}) of the converter changes.

With the goal to completely eliminate dead time related losses, it is required to control both dead times with a very high timing resolution of $\ll 500\,\text{ps}$ to regulate the dead time into the target window of Fig. 5.10a, and into the loss minimum between non-ZVS and diode conduction as depicted in Fig. 5.10b. The required dead time range can be derived from the results of Fig. 5.10b. The dead time control requires a range from $<2\,\text{ns}$ to approximately 30 ns to cover converter output currents from 50 to 500 mA.

5.4.4   Dead Time Related Losses at Light Load

The switching behavior at light load is shown in Fig. 5.11. The behavior during DT_{lo} at the falling transition of V_{sw} is identical to the typical case (Fig. 5.8a). The rising transition of V_{sw} occurs immediately after the low-side gate $V_{gs,ls}$ is turned off at the beginning of DT_{hi}, as the inductor current I_{L0} is negative (see Fig. 5.8b).

If DT_{hi} is chosen to long, as shown in Fig. 5.11a, the negative inductor current pulls up the switching node towards V_{in}. If the current is negative enough, V_{sw} even exceeds V_{in}, and the body diode of the high-side switch starts conducting. Identically to the full load case at the low-side switch, the conducting diode exhibits higher losses, than a switch in on-state. Additional body diode conduction losses occur. At the end of DT_{hi}, when the high-side switch turns on, the current commutates from the body diode to the high-side switch. Thus, the optimal turn-on time of the high-side switch is, when the rising V_{sw} exactly reaches V_{in}, a very short instant before the body diode of the high-side switch starts conducting, which is shown in Fig. 5.11b. In this case, the losses are minimal as the low-side and high-side switch are turned on with ZVS. If the dead time is chosen too short, as it is depicted in Fig. 5.11c, the high-side switch already turns on, when V_{sw} is below V_{in}. V_{sw}, and its parasitic capacitance C_{sw}, is not fully charged up lossless by the negative inductor current, but partially by the resistive on-resistance of the conducting switch. The high-side switch is not turned on with ZVS, and additional losses during each DT_{hi} occur. Nevertheless, a soft-switching condition as described in Sect. 2.2.2.4 is achieved and switching losses are significantly reduced, even with non-ideal ZVS. Similar to the loss calculation for DT_{lo} in (5.9), the dead time related losses for DT_{hi} in light load can be approximated as

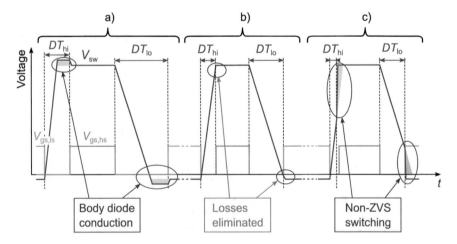

Fig. 5.11 Switching states at the switching node of the dead time controller in light-load condition (Fig. 5.8b). (**a**) Dead time too large. (**b**) Optimal dead time. (**c**) Dead time too small

$$P_{\text{DThi}} = \underbrace{\frac{1}{2} \cdot C_{\text{sw}} \cdot (V_{\text{in}} - V_{\text{sw}}(t_{\text{hs,on}}))^2 \cdot f_{\text{sw}}}_{\text{Non-ZVS Losses}} + \underbrace{I_{\text{L0}}(t_{\text{hs,on}}) \cdot t_{\text{dio,hs}} \cdot V_{\text{f}} \cdot f_{\text{sw}}}_{\text{Diode Conduction Losses}}$$

$$- \underbrace{I_{\text{L0}}(t_{\text{hs,on}})^2 \cdot R_{\text{on,hs}} \cdot t_{\text{dio,hs}} \cdot f_{\text{sw}}}_{\text{HS-On-State Losses}} . \tag{5.10}$$

In dependence, whether DT_{hi} is too short or too long, the diode conduction losses or the non-ZVS losses, respectively, become zero. $t_{\text{dio,hs}}$ is the diode conduction time of the high-side switch during DT_{hi}, which is similar to t_{dio1} in Fig. 5.8a for the low-side switch, and $V_{\text{sw}}(t_{\text{hs,on}})$ is the switching node voltage at the instant when the high-side switch turns on. $I_{\text{L0}}(t_{\text{hs,on}})$ is the inductor current, at the instant, when the high-side switch turns on. The rise time of the switching node during DT_{hi} depends strongly on the negative peak current of the inductor $I_{\text{L0}}(t_{\text{ls,off}})$ at the low-side switch turn-off, and thus on the operating point of the converter (V_{in}, I_{out} and L_0). If $I_{\text{L0}}(t_{\text{ls,off}})$ is only slightly negative, V_{sw} swings up very slowly. This significantly increases I_{L0} while V_{sw} swings up to higher input voltages, as the inductor voltage becomes positive during the swing-up. Thus, $I_{\text{L0}}(t_{\text{hs,on}})$ differs significantly from $I_{\text{L0}}(t_{\text{ls,off}})$.

To calculate the losses and to demonstrate the switching node behavior during DT_{hi}, the swing-up is simulated for different operating points. Figure 5.12a shows the behavior at $V_{\text{in}} = 12$ V for $I_{\text{out}} = 10$ mA and $I_{\text{out}} = 300$ mA. With an inductor of $L_0 = 500$ nH, both output current cases are at light load, while the negative peak inductor currents I_{Lmin} at the low-side switch turn-off are -280 and -62 mA. The small I_{Lmin} at $I_{\text{out}} = 10$ mA leads to a very strong and fast rising of V_{sw} during DT_{hi}. The switching node would significantly exceed V_{in}, assuming the high-side switch would not be turned on or its body diode would not clamp V_{sw}. As described

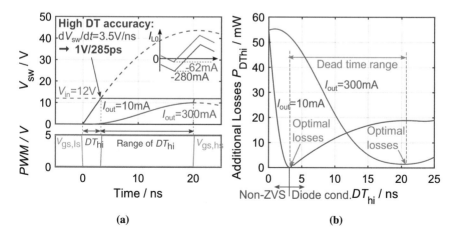

Fig. 5.12 Light-load switching conditions at $V_{\text{in}} = 12$ V, $L_0 = 500$ nH and different output currents I_{out} (10 and 300 mA); (**a**) study of the switching node swing-up and optimum high-side turn-on instant and (**b**) the dead time related losses versus the turn-high dead time DT_{hi}

in Fig. 5.11a, ZVS can be achieved and body diode conduction can be avoided if the high-side switch turns on exactly when V_{sw} reaches V_{in} before the body diode of the high-side switch starts conducting. This is challenging, as the rise time of V_{sw} is in the range of 3.5 V/ns. This means that the high-side switch needs to be turned on with a resolution of less than 285 ps to achieve ZVS with a voltage accuracy of below 1 V. For the dead time implementation, which will be shown in Sect. 6.3, a resolution of 125 ps is chosen. For the case with $I_{\text{out}} = 300$ mA (Fig. 5.12a), which is close to the border between light-load and typical load condition, the swing-up occurs only very slowly due to the small I_{Lmin} of -62 mA. V_{sw} swings up only to slightly below V_{in}. The ideal turn on is at the maximum of V_{sw}, which is at a very large dead time of $DT_{\text{hi}} = 20$ ns, at which the losses are minimum. Thus, to cover a turn-on at minimum losses, a wide dead time range of >20 ns is required. Figure 5.12b shows the additional dead time related losses for a varying dead time DT_{hi} for the load cases simulated in Fig. 5.12a. The losses are calculated according to (5.10) by extracting $I_{\text{L0}}(t_{\text{hs,on}})$ from the simulation. For $I_{\text{out}} = 10$ mA, the optimal loss point is at $DT_{\text{hi}} \approx 3$ ns. Decreasing the dead time by only a few hundred microseconds increases the losses drastically due to non-ZVS losses. For larger dead times, losses increase due to the body diode conduction of the high-side switch. The loss curve for $I_{\text{out}} = 300$ mA shows that the requirement for the dead time resolution is relaxed, but the loss optimum is shifted to $DT_{\text{hi}} \approx 20$ ns and thus requires a large dead time range, similar to the range required for DT_{lo} (compare to Fig. 5.10b).

A different operating point at $V_{\text{in}} = 48$ V is shown in Fig. 5.13a. A simulation using $L_0 = 500$ nH shows that the swing-up of V_{sw} do not reach V_{in}. However, the optimal turn-on instant is at the maximum of V_{sw}. No ideal ZVS is achieved, and the high-side switch has to discharge/charge the voltage difference of C_{sw} towards V_{in}, resulting in losses.

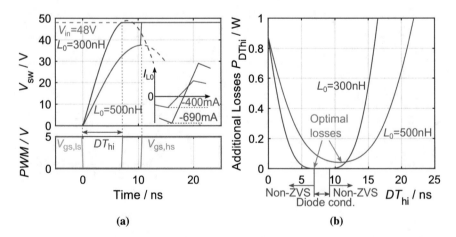

Fig. 5.13 Light-load switching conditions at $V_{\text{in}} = 48$ V, $I_{\text{out}} = 50$ mA, and two different inductors L_0 (500 and 300 nH); (a) study of the switching node swing-up and optimum high-side turn-on instant and (b) the dead time related losses versus the turn-high dead time DT_{hi}

To achieve ZVS, the inductor value has to be reduced to $L_0 = 300\,\text{nH}$, which is shown in the second operating point in Fig. 5.13a. This decreases the negative current peak at the low-side switch turn-off from -400 to $-690\,\text{mA}$ (see subset in Fig. 5.13a), and results in a stronger swing-up. If the high-side switch is turned on with ZVS at $V_{\text{sw}} = V_{\text{in}}$, the dead time related losses are theoretically eliminated. This is confirmed by calculation of the dead time related losses for varying dead times, which are depicted in Fig. 5.13b for both operating points.

However, further reduction of the inductor value significantly increases the inductor current ripple from $\Delta i_{L0} = 900\,\text{mA}$ to $\Delta i_{L0} = 1.48\,\text{A}$. Besides the on-state losses of the power switches, the AC and DC losses in the inductor significantly increase due to the higher current ripple. The gain in the reduction of the dead time related losses is thus counterbalanced by the higher inductor and on-state losses.

The best trade-off between low turn-on losses and current ripple losses is thus depending highly on the used inductor. To choose the best inductor value, it is required to study the scaling of inductor losses with current ripple and inductor value. As a study of the inductor losses was not the focus of this book, an optimized inductor value was found by experiments. Efficiency measurements on a prototype (implementation will be described in Sect. 6.3) resulted in a better efficiency using an inductor value of 500 nH for the operating point of Fig. 5.13, without achieving ideal ZVS.

The proposed dead time control is able to fully eliminate dead time related losses over a wide range of input voltage and load current variation operating in typical, as well as in light-load condition. To safely fulfill the dead time range requirements and the resolution to eliminate all dead time related losses, a dead time control is proposed in this book, which has been implemented with a nominal dead time range from 125 ps to 32 ns with a resolution of 125 ps. This enables the dead time control to regulate both DT_{hi} and DT_{lo} to loss optima for widely changing operating points. The implementation of the dead time control will be described in Chap. 6.

5.5 Loss Optimization and Limitations

If a buck converter operates at a very high conversion ratio $V_{\text{in}}/V_{\text{out}}$, the loss contribution of the high-side switch becomes very low due to the small on-time t_{on}. This allows to reduce the size of the high-side switch and thus to increase $R_{\text{on,hs}}$. A detailed calculation of the conduction losses is shown in section "Switch Conduction Losses" in the Appendix. The resulting increase of the conduction losses at the high-side switch $P_{\text{cond,hs}}$ has only a very small impact on the overall conduction losses, but parasitic capacitances of the high-side switch are reduced, and switching losses might become significantly lower.

However, increasing the on-state resistance of a switch, in order to reduce the switching losses is limited. A switch in on-state should not enter saturation region, in which its drain-source voltage and thus the on-state resistance would significantly increase, and thus the losses. For example, the PMOS switch used for the converters covered in this book is designed to have an on-state resistance of 3.4 Ω. Assuming its saturation region starts at $V_{gs} - V_{th}$, which is slightly above 4 V for $V_{gs} = 5$ V, the switch can tolerate an inductor peak current of about 1 A with a small safety margin. At a DC load current of 500 mA, a buck converter can be designed for inductor current ripples up to 100% of the DC current. The designed PMOS switch is thus also suitable as resonant switch (MPR) in the proposed parallel-resonant converter (PRC), which will be described in Sect. 7.2, in which resonant current peaks of up to 1 A are also possible in some operating conditions.

The impact of the limited on-resistance can be observed in a loss break down as shown in Fig. 2.17a; at higher switching frequencies and input voltages, the switching losses become dominant compared to the static on-state resistance losses.

The loss break down of an asynchronous buck converter shown in Fig. 5.4a demonstrates that the used Schottky diode (SS16) has a significant impact on converter losses. The diode [13] is typically used in automotive applications, which require the same current and voltage range as covered in this book. The availability of power diodes with a more optimized capacitance and forward voltage is limited on the market, especially for diodes with high break down voltages and low currents. Further improvements of the Schottky diode towards the requirement of fast-switching high-V_{in} converters could have a high potential to further reduce switching losses and improve power efficiency of the asynchronous buck converter.

5.6   Architecture Comparison

To determine, which buck converter architecture is superior in power efficiency, a transient transistor level simulation of a power stage implementation for each architecture is done to compare the efficiency. To only compare the power stage, an ideal inductor was used to neglect inductor losses. The inductor value was chosen large enough to have a negligibly small inductor current ripple. In the asynchronous buck converter, the NMOS transistor was designed to have approximately the same size, resulting the parasitic capacitances to be in the same range. The NMOS switch was implemented as a lateral diffusion MOS (LDMOS) transistor, while the PMOS switch was only available as drain-extended transistor in the used technology. This results in a significantly better on-state resistance for the NMOS switch ($R_{on} = 0.8 \Omega$) compared to the PMOS switch ($R_{on} = 3.4 \Omega$). In the synchronous converter, both the high-side and the low-side switch were implemented identically (each with $R_{on} = 0.8 \Omega$).

Figure 5.14a shows a first comparison of an asynchronous buck converter power stage once with an NMOS switch, and once with a PMOS switch, both operating at 10 MHz, $V_{out} = 5$ V, and a medium input voltage of $V_{in} = 18$ V. It can be observed that at lower output currents, the difference in efficiency is marginal, while at higher output currents, the synchronous buck converter becomes superior (up to 6% better), as the on-state resistance of the switch becomes dominant in the loss contribution. Figure 5.14a also shows two simulations for each converter architecture, one is

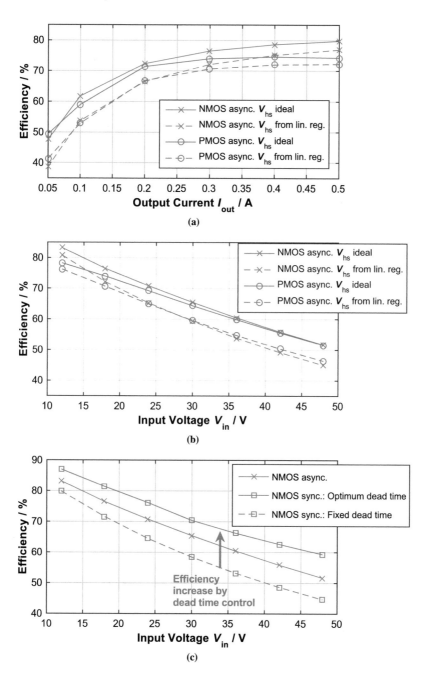

Fig. 5.14 Efficiency comparison of different buck converter architectures. (**a**) Comparison of the power efficiency of an asynchronous buck converter with NMOS and PMOS power switch versus the output current, operating at 10 MHz, $V_{in} = 18$ V and $V_{out} = 5$ V. (**b**) Comparison of the power efficiency of an asynchronous buck converter with NMOS and PMOS power switch versus the input voltage, operating at 10 MHz, $I_{out} = 0.3$ A and $V_{out} = 5$ V. (**c**) Comparison of the power efficiency of an asynchronous to a synchronous buck converter with optimized and non-optimized dead time versus the input voltage, operating at $f_{sw} = 10$ MHz, $I_{out} = 0.3$ A and $V_{out} = 5$ V

assumed to have an ideal high-side supply (i.e. connecting an ideal source between V_{boot} and $HSGND$, see also Fig. 3.2), the second one assumes that the high-side supply is generated by a linear regulator as introduced in Fig. 3.3a for the PMOS switch. For the NMOS switch, it is assumed that the input supply of the bootstrap circuit, used to generate the high-side supply (V_{dd5} in Fig. 3.1c) is generated from V_{in} by a linear regulator. The losses caused by this linear regulator do not change with the output current, consequently it leads to a significantly dropping efficiency at lower output currents, as the losses related to the output power are less.

In Fig. 5.14b, this comparison is done for a varying input voltage V_{in} with $I_{out} = 300$ mA. In this case, the difference in efficiency becomes small at high V_{in}. As the duty cycle decreases for higher input voltages, the on-time of the high-side switch becomes very small. The shorter conduction time (on-time) of the high-side switch also reduces the on-state losses. At large duty cycles at low input voltages, the on-state losses become dominant again, resulting in the NMOS switch being superior to the PMOS switch. Figure 5.14b again compares the efficiency with an ideal high-side supply to a high-side supply generated by linear regulators for both NMOS and PMOS switch. As the voltage drop over the pass device MP_{lin} of the linear regulator is $HSGND = V_{in} - 5$ V (see Fig. 3.3a) for a PMOS switch, or $V_{in} - V_{dd5}$ (with V_{dd5} as input of the bootstrap circuit as shown in Fig. 3.3c), the losses decrease at low V_{in}, while at high V_{in}, the efficiency difference to an ideal high-side source is significant (up to 7%).

For a decision to use an NMOS or PMOS switch in an asynchronous buck converter following arguments can be derived.

• For a large operating range, the difference in efficiency is not significant. If the converter is mainly operating at its maximum load and at the same time at low or moderate conversion ratios (large duty cycles), an NMOS is superior. If this is not the case, a PMOS might be the preferred choice as its implementation is less critical, with respect to the level shifter design and substrate coupling, as the high-side supply is not switching (compare Fig. 3.2 and Sects. 3.4 and 4.3).
• The implementation of the generation of the high-side supply has a large impact on the efficiency. For the PMOS switch, the high-side supply could be generated with a charge pump as shown in Fig. 3.3b. However, it requires an additional switching driver connected to the pumping capacitor, which needs to deliver a charge which is higher than the gate charge of the PMOS high-side switch, and creates additional losses. It improves the efficiency compared to a linear regulator, but an efficiency close to an ideal high-side supply is hardly reached. In contrast, the high-side supply can be generated by a bootstrap circuit (see Fig. 3.1c), which is driven by the switching node. No additional driver is required which makes a bootstrap circuit efficient. However, the supply of the bootstrap circuit has to be generated efficiently. In larger power management ICs, typically a separate high-efficient converter is implemented to generate the low-side supplies for several converters. This can be used as an efficient supply for the bootstrap circuit. A second option is to power up the buck converter with a linear regulator supplying the bootstrap circuit, and connecting the output voltage to the bootstrap circuit as soon as it is stable. This is only possible if the output voltage

is in the range of the required high-side supply voltage. If such a high-efficient supply for the bootstrap circuit is available on the system, the converter with an NMOS switch will likely be more efficient as a PMOS device.

- The efficiency of an asynchronous buck converter also depends significantly on the available PMOS and NMOS switch technology and their figure of merits related to parasitic capacitances and on-state resistance, as demonstrated in Sect. 3.2.

Figure 5.14c compares an asynchronous buck converter with NMOS switch to a synchronous converter, with both high-side and low-side switch implemented as NMOS transistors. For the synchronous converter, two simulations are shown, one with a dead time adjusted to its optimum value at each operating point, and a second one, with fixed non-optimal dead times of $DT_{hi} = 5\,ns$ and $DT_{lo} = 1\,ns$ across all operating points. The values are chosen based on a realistic scenario, in which the dead time generator adjusts both dead times to 3 ns, and the level shifter delays the high-side signal by 2 ns. The level shifter delay adds up to DT_{hi} and subtracts from DT_{lo}, resulting in $DT_{hi} = 5\,ns$ and $DT_{lo} = 1\,ns$.

From the efficiency comparison of Fig. 5.14c, it can be observed that dead time drastically impacts the efficiency by more than 15%. Comparing the efficiency from the asynchronous converter to the synchronous converter, it can be concluded that a synchronous converter can be superior to an asynchronous converter (by up to 10%), only when the dead time is adjusted to its optimal value at each operating point, which matches the theoretical analysis of dead time related losses in Sect. 5.4.3.2.

An asynchronous buck converter architecture leads to the highest efficiency, but a highly accurate dead time control adjusting dynamically with changing operating points is required. In Sect. 6, a dead time control implementation is proposed which is able to fully eliminate dead time related losses.

5.7 Design Indicator: Efficiency Scaling

A major challenge is to compare and predict the efficiency at different operating points. Due to the scaling of the losses and the efficiency with varying parameters, like input and output voltage, switching frequency and load, the expected efficiency is often changing significantly over different operating points.

The efficiency η scales in dependency of the output voltage V_{out}, the output current I_{out}, and thus, the output power $P_{out} = V_{out} \cdot I_{out}$, as well as the power losses P_{loss} of the converter, and is defined as

$$\eta = \frac{V_{out} \cdot I_{out}}{V_{out} \cdot I_{out} + P_{loss}(f_{sw}, V_{in}, I_{out})} \tag{5.11}$$

$$= \frac{1}{1 + \dfrac{P_{loss}(f_{sw}, V_{in}, I_{out})}{V_{out} \cdot I_{out}}} \tag{5.12}$$

$$= \frac{1}{1 + \underbrace{P_{loss,norm}(f_{sw},\, V_{in},\, I_{out}V_{out})}_{\text{Normalized losses}}}. \tag{5.13}$$

Equation (5.11) can be re-arranged to (5.12) in order to obtain a fraction including the losses P_{loss} and the output power $V_{out} \cdot I_{out}$. The losses P_{loss} in (5.12) scale with changing f_{sw}, V_{in}, I_{out}, and V_{out}. As soon as the switching losses in a converter are dominant, the losses scale approximately proportionally to f_{sw} and in a first order to V_{in}^2. In the following, it is assumed that the losses also scale proportional with I_{out}, which is the case for the assumption that the power switches of two converters, operating at different output currents, are scaled such that the on-state resistance is inversely proportional to I_{out} to balance on-state and switching losses.

In the next step, the losses P_{loss} are normalized to the output power, resulting in the normalized losses P_{loss0} as shown in (5.13). Assuming now that a converter operating point changes, a given efficiency η scale with the normalized losses $P_{loss,norm}$, as a function of f_{sw}, V_{in}, V_{out}, and I_{out}.

With f_{sw0}, V_{in0}, and I_{out0} being the values of an initial reference operating point, and f_{sw}, V_{in}, and I_{out} being the values at a different second operating point, the scaling of the converter losses from the initial to the second operating point can be expressed as

$$\underbrace{\frac{P_{loss0}}{f_{sw0} \cdot V_{in0}^2 \cdot I_{out0}}}_{\text{Initial operating point}} \approx \underbrace{\frac{P_{loss}}{f_{sw} \cdot V_{in}^2 \cdot I_{out}}}_{\text{Second operating point}}. \tag{5.14}$$

Equation (5.14) can be re-arranged to

$$P_{loss} \approx \underbrace{\frac{P_{loss0}}{f_{sw0} \cdot V_{in0}^2 \cdot I_{out0}}}_{\text{Initial operating point}} \cdot \underbrace{f_{sw} \cdot V_{in}^2 \cdot I_{out}}_{\text{Loss scaling factor}}. \tag{5.15}$$

The normalized losses $P_{loss,norm}$ from (5.12) and (5.13) are defined as

$$P_{loss,norm} = \frac{P_{loss}}{V_{out} \cdot I_{out}}. \tag{5.16}$$

Thus, $P_{loss,norm}$ scale with V_{out} and I_{out}. Substituting P_{loss} in (5.16) with (5.15) leads to the overall scaling of the normalized losses, and thus of the converter efficiency. The scaling of the normalized losses is obtained as

$$P_{loss,norm} \approx \underbrace{\frac{P_{loss0}}{f_{sw0} \cdot V_{in0}^2 \cdot I_{out0}}}_{\text{Initial operating point}} \cdot \underbrace{\frac{f_{sw} \cdot V_{in}^2 \cdot I_{out}}{V_{out} \cdot I_{out}}}_{\text{Loss scaling factor = Design indicator}}. \tag{5.17}$$

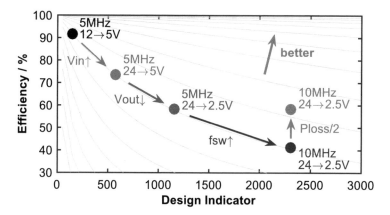

Fig. 5.15 Efficiency scaling versus the design indicator for various operating points

A design indicator is proposed, which is equal to the resulting loss scaling factor of the normalized losses $P_{\text{loss,norm}}$. The design indicator derived from the loss scaling factor of (5.17) is thus defined as

$$DI = \frac{f_{\text{sw}} \cdot V_{\text{in}}^2}{V_{\text{out}}}. \tag{5.18}$$

The proposed design indicator is a measure for the scaling of the normalized losses $P_{\text{loss,norm}}$, and thus of the overall converter efficiency, which allows to compare two different operating points.

Based on (5.18), scaling of the efficiency can be drawn as a function of the design indicator, as shown in Fig. 5.15. By plotting the efficiency versus the normalized losses $P_{\text{loss,norm}}$ as calculated in (5.13), efficiency contour lines are obtained. Different initial operating points P_{loss0}, f_{sw0}, V_{in0}, V_{out0}, and I_{out0}, resulting in different initial normalized losses lead to different contour lines. A change of any parameter f_{sw}, V_{in}, or V_{out} scales the initial efficiency along the efficiency contour line. The resulting distance on the contour line to the original efficiency point is defined by the scaling factor of each parameter in the design indicator, according to (5.18). This is demonstrated in Fig. 5.15 for operating points with increasing V_{in}, decreasing V_{out}, and increasing f_{sw}. I_{out} do not scale the efficiency as demonstrated in (5.17) and (5.18). For example, the efficiency of a 5 MHz, 12 V-to-5 V converter scales down from above 90% to about 73% if the input voltage is doubled. A further reduction to half of V_{out} reduces the efficiency to around 60%. If the switching frequency is doubled, the efficiency drops to below 45%. Assuming that the converter performance is improved to only have half the switching losses, for example by applying soft-switching techniques, the converter operating at $V_{\text{in}} = 24$ V, $V_{\text{out}} = 2.5$ V, and $f_{\text{sw}} = 10$ MHz shifts up to a higher efficiency contour line, indicating the improved converter performance. Following the efficiency contour line of the improved converter, the improved converter can be directly compared to the original converter at a completely different operating point, for example at 5 MHz, 12 V input, and 5 V output, which is located at a lower efficiency contour line.

The main benefit of the design indicator is the possibility to compare converters published at different operating points. This will be demonstrated in Sect. 7.2, which benchmarks the converters of this book against published state-of-the-art converters.

An efficiency comparison using the design indicator is limited by assuming that the switching losses are dominant in the overall loss scaling, while the influence of the static losses is neglected. Therefore, the design indicator allows only an approximate comparison. A more accurate definition by including the static losses in the overall loss scaling would make the design indicator too complex for practical use. Moreover, the information provided in publications are typically limited to the parameters used for the proposed design indicator. Thus, a more accurate design indicator is not suitable to compare published converters due to the lack of information. As previously no tool was available, and thus published converters could practically not be compared, the proposed design indicator delivers a significantly improved quality of comparison.

Appendix

Switch Conduction Losses

The total switch conduction losses P_{cond} in buck converter are the sum of the conduction losses $P_{cond,hs}$ of the high-side switch and $P_{cond,ls}$ of the low-side switch.

Average losses over a power switch (conduction losses) in on-state during the on-time t_{on} with an on-state resistance $R_{on,hs}$ of the high-side switch are calculated as

$$\overline{P_{cond,hs}} = \int_{0}^{t_{on}} R_{on,hs} \cdot I_{L0}^{2}(t). \tag{5.19}$$

During the on-time of, e.g., the high-side switch, the inductor current is linearly rising from $I_{Lmin} = I_{out} - \Delta i_{L0}/2$ to $I_{Lmax} = I_{out} + \Delta i_{L0}/2$ (current ramp). In this time, the inductor current is

$$I_{L0}(t) = I_{Lmin} + \frac{\Delta i_{L0}}{t_{on}} \cdot (t). \tag{5.20}$$

Inserting (5.20) into (5.19) and solving the integral results in

$$\overline{P_{cond,hs}} = R_{on,hs} \left(I_{Lmin}^{2} + I_{Lmin} \cdot \Delta i_{L0} + \frac{1}{3} \Delta i_{L0}^{2} \right). \tag{5.21}$$

Expressing (5.21) by the output current by substituting $I_{Lmin} = I_{out} - \Delta i_{L0}/2$ leads to

$$\overline{P_{cond,hs}} = R_{on,hs} \left(I_{out}^2 + \frac{1}{12} \Delta i_{L0}^2 \right). \tag{5.22}$$

As t_{on} is vanishing in (5.21), the average power for a ramp current is independent of the ramp time. Consequently, the calculation of $P_{cond,ls}$ is accordingly for the falling current ramp from I_{Lmax} to I_{Lmin} during the on-time of the low-side switch, which is t_{off} of the buck converter, with the on-state switch resistance $R_{on,ls}$.

Thus, the total conduction losses are

$$\overline{P_{cond}} = \frac{1}{T} \left(\overline{P_{cond,hs}} \cdot t_{on} + \overline{P_{cond,ls}} \cdot t_{off} \right)$$

$$= \left(R_{on,hs} \cdot \frac{t_{on}}{T} + R_{on,ls} \cdot \frac{t_{off}}{T} \right) \cdot \left(I_{out}^2 + \frac{1}{12} \Delta i_{L0}^2 \right). \tag{5.23}$$

Writing the equation in dependence of V_{in} and V_{out} by substituting the duty cycle $t_{on}/T = V_{out}/V_{in}$ results in

$$\overline{P_{cond}} = \left(R_{on,hs} \cdot \frac{V_{out}}{V_{in}} + R_{on,ls} \cdot \left(1 - \frac{V_{out}}{V_{in}} \right) \right) \cdot \left(I_{out}^2 + \frac{1}{12} \Delta i_{L0}^2 \right). \tag{5.24}$$

The first term of (5.24) shows that the losses of the high-side switches contribute with a factor of $R_{on,hs} \cdot V_{out}/V_{in}$, while the losses of the low-side switch contribute with a factor of $R_{on,ls} \cdot (1 - V_{out}/V_{in})$.

Calculation of the DC Losses in the Inductor

The inductor DC losses $P_{L,DC}$ are calculated the same way as the switch conduction losses, as the inductor current ramps are also causing losses at the inductor DC resistance R_{dcr}, instead of the switch on-state resistance. The average power loss $P_{L,DC}$ are thus calculated by modifying (5.23) (substitution of $R_{on,hs}$ and $R_{on,ls}$) to

$$\overline{P_{L,DC}} = \left(R_{dcr} \cdot \frac{t_{on}}{T} + R_{dcr} \cdot \frac{t_{off}}{T} \right) \cdot \left(I_{out}^2 + \frac{1}{12} \Delta i_{L0}^2 \right)$$

$$= R_{dcr} \cdot \left(I_{out}^2 + \frac{1}{12} \Delta i_{L0}^2 \right). \tag{5.25}$$

References

1. Basso C (2008) Switch-mode power supplies spice simulations and practical designs, 1st edn. McGraw-Hill, New York
2. Benda H, Spenke E (1967) Reverse recovery processes in silicon power rectifiers. Proc IEEE 55(8):1331–1354. https://doi.org/10.1109/PROC.1967.5834
3. Chinag CY, Chen CL (2009) Zero-voltage-switching control for a PWM buck converter under DCM/CCM boundary. IEEE Trans Power Electron 24:2120–2126. https://doi.org/10.1109/TPEL.2009.2021186
4. Eberle W, Zhang Z, Liu YF, Sen PC (2009) A practical switching loss model for buck voltage regulators. IEEE Trans Power Electron 24:700–713. https://doi.org/10.1109/TPEL.2008.2007845. http://ieeexplore.ieee.org/document/4757277/
5. Graovac DD, Pürschel M, Kiep A (2006) MOSFET power losses calculation using the datasheet parameters. Application note, Infineon, Automotive Power
6. Kelin J (2006) Synchronous buck MOSFET loss calculations with excel model. https://www.fairchildsemi.com/application-notes/AN/AN-6005.pdf
7. Li CH, Lo YK, Chiu HJ, Chen TY (2012) Accurate power-loss estimation for continuous-current-conduction-mode synchronous buck converters. In: Anti-counterfeiting, security, and identification, pp 1–5. https://doi.org/10.1109/ICASID.2012.6325298. http://ieeexplore.ieee.org/document/6325298/
8. Meade T, O'Sullivan D, Foley R, Achimescu C, Egan M, McCloskey P (2008) Parasitic inductance effect on switching losses for a high frequency DC-DC converter. In: 2008 Twenty-third annual IEEE applied power electronics conference and exposition, pp 3–9. https://doi.org/10.1109/APEC.2008.4522692. http://ieeexplore.ieee.org/document/4522692/
9. Orabi M, Shawky A (2015) Proposed switching losses model for integrated point-of-load synchronous buck converters. IEEE Trans Power Electron 30:5136–5150. https://doi.org/10.1109/TPEL.2014.2363760. http://ieeexplore.ieee.org/document/6926802/
10. Orabi M, Abou-Alfotouh A, Lotfi A (2008) Coss capacitance contribution to synchronous buck converter losses. In: 2008 IEEE power electronics specialists conference, pp 666–672. https://doi.org/10.1109/PESC.2008.4592006. http://ieeexplore.ieee.org/xpl/articleDetails.jsp?arnumber=4592006
11. Pam S, Sheehan R, Mukhopadhyay S (2012) Accurate loss model for DC-DC buck converter including non-linear driver output characteristics. In: 2012 Twenty-seventh annual IEEE applied power electronics conference and exposition (APEC), pp 721–726. https://doi.org/10.1109/APEC.2012.6165899. http://ieeexplore.ieee.org/document/6165899/
12. Ren Y, Xu M, Zhou J, Lee FC (2005) Analytical loss model of power MOSFET. IEEE Trans Power Electron 21:310–319. https://doi.org/10.1109/TPEL.2005.869743. http://ieeexplore.ieee.org/xpl/articleDetails.jsp?arnumber=1603662
13. Vishay (revision 21-Jul-2017) Surface mount Schottky barrier rectifier. SS16 Datasheet, Document Number: 88746
14. Wang X, Huang AQ (2011) Capacitor energy variation based designer-side switching losses analysis for integrated synchronous buck converters in CMOS technology. https://doi.org/10.1109/APEC.2011.5744736. http://ieeexplore.ieee.org/xpl/articleDetails.jsp?arnumber=5744736
15. Wang X, Huang AQ (2011) Considerations on the optimal power stage segmentation algorithm for MHz integrated synchronous Buck DC-DC converters. https://doi.org/10.1109/ISPSD.2011.5890821. http://ieeexplore.ieee.org/document/5890821/
16. Wang X, Park J, Brunt ERV, Huang AQ (2010) Switching losses analysis in MHz integrated synchronous buck converter to support optimal power stage width segmentation in CMOS technology. https://doi.org/10.1109/ECCE.2010.5618055. http://ieeexplore.ieee.org/document/5618055/

17. Wang J, Chung HS, Li RT (2013) Characterization and experimental assessment of the effects of parasitic elements on the MOSFET switching performance. IEEE Trans Power Electron 28:573–590. https://doi.org/10.1109/TPEL.2012.2195332. http://ieeexplore.ieee.org/document/6192346/
18. Wittmann J, Barner A, Rosahl T, Wicht B (2015) A 12V 10MHz buck converter with dead time control based on a 125 ps differential delay chain. In: European solid-state circuits conference (ESSCIRC), ESSCIRC 2015 – 41st, pp 184–187. https://doi.org/10.1109/ESSCIRC.2015.7313859
19. Wittmann J, Barner A, Rosahl T, Wicht B (2016) An 18V input 10MHz buck converter with 125ps mixed-signal dead time control. IEEE J Solid-State Circuits 51(7):1705–1715. https://doi.org/10.1109/JSSC.2016.2550498
20. Yan W, Pi C, Li W, Liu R (2010) Dynamic dead-time controller for synchronous buck DC-DC converters. Electron Lett 46(2):164–165. https://doi.org/10.1049/el.2010.2651

Chapter 6
Dead Time Control

As described in Sects. 5.4 and 5.6, a high-resolution dead time control is required for a synchronous converter architecture to be superior in power efficiency. In this book, a dead time control implementation for both typical operation (Fig. 5.8a) and light-load condition (Fig. 5.8b) is proposed, which is able to fully eliminate dead time related losses along varying operating points. This is achieved if the turn-high and the turn-low dead time are chosen to have each the required resolution of 125 ps a range of 32 ns, as it was derived in Sect. 5.6. The implementation of the dead time control for full-load condition is shown in Sect. 6.2, while Sect. 6.3 explains how the implemented dead time control can be extended to operate in light-load condition to achieve soft-switching of the low-side and high-side switch.

6.1 State-of-the-Art Dead Time Controls

Several dead time control concepts were published for lower switching frequencies. Most state-of-the-art concepts evaluate the switching condition by measuring the body diode forward voltage at the low-side switch. Adaptive dead time control methods instantly turn on the high-side switch as soon as the body diode starts conducting. Due to the propagation delay of the detection circuits, the achieved minimum dead time is limited to tens of nanoseconds [6], especially in high-voltage technologies, which are limited in bandwidth and speed. This is circumvented by predictive dead time control, which adjusts the dead time for the subsequent period [6]. Maximum Power Point Tracking (MPPT) concepts evaluate the converter efficiency over several periods, slowly adjusting a new dead time to a new operating point [1, 11]. The concept reported in [5] for switching frequencies up to 10 MHz achieves a dead time resolution of 1 ns by sensing the currents in the power switches

© Springer Nature Switzerland AG 2020
J. Wittmann, *Integrated High-V$_{in}$ Multi-MHz Converters*,
https://doi.org/10.1007/978-3-030-25257-1_6

and detecting the body diode conduction. However, body diode conduction losses cannot be eliminated completely, as a conducting diode is required to detect the dead time condition.

6.2 Predictive Mixed-Signal Dead Time Control

This book proposes a high-resolution predictive dead time control, suitable for switching frequencies of $f_{sw} > 10\,\text{MHz}$ and input voltages of $V_{in} > 18\,\text{V}$. The dead time does not depend on the body diode forward conduction, and thus allows a complete elimination of dead time related losses.

The dead times DT_{hi} and DT_{lo} (Fig. 5.8a) can be adjusted by changing either the high-side turn-on and turn-off instants $t_{hs,on}$ and $t_{ls,off}$, or by changing the low-side turn-on and turn-off instants $t_{ls,off}$ and $t_{ls,on}$. Both dead times have to be adjusted independently. Modifying $t_{hs,on}$ and $t_{ls,off}$ would change the V_{sw} transition instant with significant impact on the on-time of the converter and, thus, the duty cycle as well as the output voltage, which is the averaged switching node voltage V_{sw}. Therefore, the proposed dead time control has been chosen to control $t_{ls,off}$ and $t_{ls,on}$ to adjust both dead times DT_{hi} and DT_{lo}. This way, the output voltage receives only a slight change if the durations of the diode conduction t_{dio1} and t_{dio2} (Fig. 5.8a) are adjusted by dead time regulation. The slight change comes from the fact that V_{sw} is lower during body diode conduction ($V_{sw} = -V_f$) compared to the time, when the low-side switch is turned on ($V_{sw} = -V_{on}$).

A change in the converter's duty cycle will be corrected by the output voltage regulation. As soon as the output voltage and the dead time control reach a steady state, a change in the duty cycle is undesired. Therefore, a fine-adjustment of the dead times for only slightly modifying operation conditions should happen with only very small changes in the dead times, which in addition justifies a high time resolution of the digitally controlled dead times.

6.2.1 Dead Time Control Concept

The control technique proposed in this book uses a predictive algorithm to control both dead times DT_{hi} and DT_{lo}. Based on the three cases, outlined in Fig. 6.1, the dead time setting can be evaluated. The dead times can be either too long, too short or optimal. The status of DT_{hi} is obtained by sampling the switching node a few hundred picoseconds (one inverter delay) after the low-side switch is turned off at sample point S_1. At a too long dead time, V_{sw} is pulled low due to body diode conduction. S_1 is between $-V_{on}$ and $-V_f$ (Fig. 6.1a). If DT_{hi} is too short, i.e., cross conduction occurs, the high-side switch is already turned on when the low-side switch turns off, and the sample point at S_1 is at a positive voltage (Fig. 6.1c).

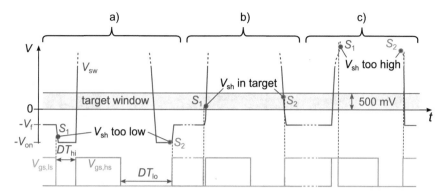

Fig. 6.1 Sampling and regulation principle of the dead time controller. (**a**) Dead time too large. (**b**) Optimal dead time. (**c**) Dead time too small

The dead time is considered optimal if the sampled voltage at S_1 is within a defined target window with a range from 0 to 500 mV. With a maximum V_{ds} of 500 mV, the low-side switch is deeply in triode region with limited drain current. As this happens during a very short time (tens of picoseconds), a 500 mV window ensures that there is effectively no cross conduction between the high-side and low-side switches.

This assures that the low-side switch is turned off until the drain-source voltage of the conducting low-side switch exceeds 500 mV. Thus, the low-side switch only conducts deeply in linear region, which limits cross currents, even when the high-side switch already starts to turn on. To evaluate the dead time DT_{lo}, a second sampling is performed at S_2, which is triggered by the signal of the last low-side gate driver stage (see signal V_n in Fig. 4.12, Sect. 4.4), which occurs a few hundred picoseconds (delay of one driver stage) before the low-side switch turns on. For a dead time, which is too long, the diode was conducting before the low-side turns on. This results again in a sampled voltage around $-V_f$. If the dead time is too short, the low-side switch is turned on at a high switching node voltage V_{sw} (non-ZVS), which leads to a positive sampled voltage at S_2. DT_{lo} is considered optimal if the sampled voltage at S_2 is in the target window between zero and 500 mV. This ensures that the error towards ZVS is always less than 500 mV, resulting in a negligible impact on efficiency.

The implementation of the proposed dead time control concept is shown in Fig. 6.2. The PWM and the inverted PWM signals are each passed through an 8-bit digital delay chain. DT_{hi} is generated by delaying the rising edge of the PWM signal and DT_{lo} by delaying the falling edge of the original PWM signal. A digital multiplexer generates $CTRL_{ls}$ and $CTRL_{hs}$ out of the delayed PWM signals. Each delay line is controlled by two digital up-down counters, which increase or decrease if the sampled voltages at S_1 or S_2 (Fig. 6.1) are below the target window at too large dead times, or above the target window at too small dead times. The evaluation is done by window comparators, comparing the sampled voltages to the target window.

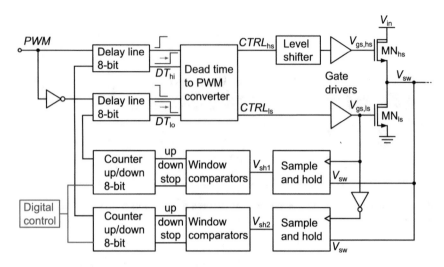

Fig. 6.2 Detailed overview of the proposed mixed-signal dead time control

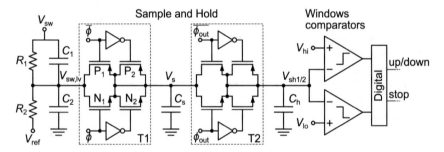

Fig. 6.3 Frequency compensated charge balanced sample and hold circuit with window comparators

At optimal dead times, if S_1 or S_2 are within the target window, the counters are stopped, and a steady state is achieved. S_1 and S_2 are stored in a sample and hold circuit.

6.2.2   Sample and Hold Circuit

The sampling is done by a high-speed sample and hold circuit, shown in Fig. 6.3 with the timing depicted in Fig. 6.4. By matching the impedance of C_1, R_1, and C_2, R_2, a frequency compensated voltage divider scales down the switching node voltage into the low-voltage 5 V-supply domain. A scaling of $V_{sw,lv} = 1/10 \cdot V_{sw}$ is achieved with sufficient accuracy in the picoseconds range. This enables input voltages up to 50 V. The divider also scales down the 500 mV target window (Fig. 6.1) to approximately 50 mV at $V_{sw,lv}$.

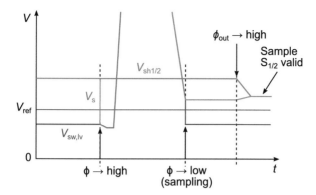

Fig. 6.4 Signals of the sample and hold circuit

To sample negative voltages at V_{sw} and to keep the divided signal level above zero, an offset is added to $V_{sw,lv}$ by connecting R_2 to a positive reference $V_{ref} = 0.8$ V. This avoids leakage currents, as the transfer gate T1 would not turn off for $V_{sw} < 0$. $\overline{\Phi}$ and Φ are controlled such that T1 is turned off at the sampling events S_1 and S_2 (Fig. 6.1), and the divided switching node voltage $V_{sw,lv}$ is stored on the sampling capacitance C_s. A charge compensation for the transfer gate is achieved by N2 and P2, as transistors N1 and P1 would couple a part of its gate charge into the hold capacitance C_s.

A second transfer gate T2 is switched on for a few nanoseconds after T1 is safely disconnected. This transfers the voltage V_s to $V_{sh1/2}$ and stored onto the hold capacitor C_h. To settle the output voltage within the on-time of the second transfer gate, C_h must be much smaller than C_s. This design uses $C_h = 50$ fF (including the input capacitance of the following comparator), $C_s = C_1 = 250$ fF, and $C_2 = 2$ pF. The sampled voltage $V_{sh1/2}$ at C_h is present during the entire switching period of the converter, while C_h is connected back again to the input divider (Fig. 6.4). This way $V_{sh1/2}$ is only slowly varying as long as the operating point of the converter does not change.

The subsequent window comparators toggle only if $V_{sh1/2}$ crosses the border of the target window in the upper or lower direction. The comparators have low timing requirements, as the comparator outputs only change the counting direction of the counter (Fig. 6.2), or stop the counter in the target window when the operating point of the converter is reached. A conventional 2-stage miller amplifier with high gain is sufficient to perform the comparison within one switching period. The target window references V_{hi} and V_{lo} are derived from a bandgap reference. They are trimmable, such that the level of the window as well as the window width can be adjusted. In the implemented dead time control, a width of 50 mV, resulting in 500 mV up-scaled target window at V_{sw}, is selected, as the mismatch variation of each window comparator has an untrimmed 3σ input offset of $\sim$20 mV. This assures that the comparator levels do not overlap, i.e., the window width is always positive ($\geq$10 mV), but the window is kept small enough to not cause significant dead time related losses. In addition, the sampling stage is placed very close to the integrated power switches to ensure minimum influence from bond or board parasitics during switching.

6.2.3 Differential Delay Lines

The dead times are generated by 8-bit digitally adjustable delay elements with a nominal dead time resolution of 125 ps. At 8-bit, a dead time range of 32 ns for each dead time DT_{hi} and DT_{lo} is obtained to cover a wide output current range (in line with Sect. 5.4.3, Fig. 5.10). The high resolution is achieved by a differential dead time generation, shown in Fig. 6.5a [10]. In each LSB element, two parallel inverter chains are installed. A delay capacitance C_{del} in one of the two branches increases the propagation delay slightly with respect to the other branch. To be able to use a digital design flow, standard inverter gates are used to realize the capacitance C_{del}. The difference of the propagation delay between the two branches of the differential delay chain is utilized to generate the dead time. The LSB delay elements are connected in series. The differential delay is set by tapping the appropriate stage

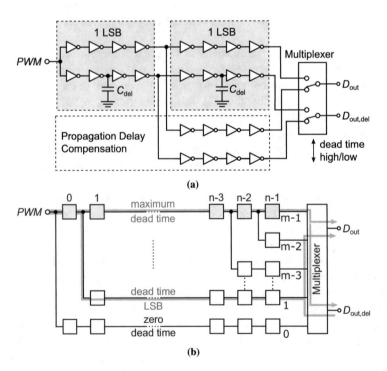

Fig. 6.5 Implementation of the digital delay line, **(a)** with two differential LSB delay elements, and **(b)** with n delay elements

output of the delay chain by a digital multiplexer. This way, the adjustable dead time can be smaller than the technology related minimum propagation delay of two inverters, which are required to preserve the signal polarity. In the proposed design, the implementation of one LSB with a differential delay of nominal 125 ps is less than half of the minimum possible delay of two minimal inverters. Similar to the inverter delay, the differential delay varies due to process variations by approximately $\pm 30\%$, which has to be considered in the overall design of the delay chains. Theoretically, the minimum value of the differential delay is only limited by the inverter's random delay mismatch of a few picoseconds as there is no minimum limit to C_{del}. C_{del} could even be implemented as pure wiring capacitance.

The time difference between D_{out} and $D_{out,del}$ is small, but each of the signals has a large absolute propagation delay compared to the PWM signal. Selecting different delays, only the differential delay, and thus the dead time, should change. To avoid a time step of the overall PWM signal, the propagation delay of the LSB delay elements, which are bypassed by the multiplexer, has to be compensated as shown in Fig. 6.5b. Both signals D_{out} and $D_{out,del}$ are passed through a propagation delay compensation element, consisting of a pure inverter stage for each signal, without adding a differential delay. For the longest dead time setting at the maximum counter value, all n differential delay elements would be selected (Fig. 6.5b). At a counter value of 1, for example, one differential delay chain and $n-1$ compensation elements would be used, while at zero dead time, n compensation elements and no delay element are required. This way, all dead time settings have the same overall propagation delay.

To reduce the number of sub-cells and the absolute propagation delay, the 8-bit delay chain is divided into two different 4-bit delay chains as shown in Fig. 6.6, the first with 125 ps per LSB, and the following stage with $16 \times 125\,\text{ps} = 2\,\text{ns}$ per LSB with a 16× larger C_{del}. The cascaded delay chains reduce the number of delay elements n and multiplexed delay paths m to $n = m = 16$ in each 4-bit delay chain. In this design, the 125 ps delay element is 20% larger than the compensation element, and the 2 ns delay element has approximately three times the size of the 125 ps delay element.

As the sampled voltage is regulated to the target window for the optimal dead time, only process and temperature of the sampling stage and the accuracy of the window comparators have an influence on the accuracy of the dead time. Variations in the sampling occur between the sampling signal at the transfer gate T1 and the actual low-side switch turn-on, which is in the range of few hundred picoseconds. Large variations of the mismatch between the propagation delay from the delay

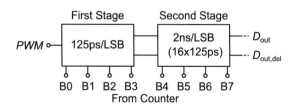

Fig. 6.6 Area efficient implementation of the 8-bit delay line by two cascaded 4-bit stages

chains to the low-side switch and from the delay lines to the high-side switch, respectively, are practically entirely compensated by the dead time control. The mismatch, generated by an asymmetry between the low-side and high-side circuits, e.g., due to the level shifter, can be in the range of several nanoseconds, which have to be considered in the definition of the dead time ranges.

While the range of DT_{hi} has to cover mainly the process variations, the optimum dead time DT_{lo} also scales up with the input voltage, and is inversely proportional to the output current, which discharges V_{sw} during DT_{lo} (see Sect. 5.4.3, Fig. 5.10b). In general, dead time regulation is possible even beyond the specification of this design, i.e., for $V_{in} > 18$ V, $I_{out} < 50$ mA. This requires to increase the number of delay elements in the second delay chain (Fig. 6.6). One additional bit in the second delay chain yields twice the dead time range and results in a $2.5\times$ area increase of the entire delay line in Fig. 6.6. An operation at significantly higher output current would require a larger size of the power switches to not increase on-state losses. This scales up the parasitic capacitances, which compensates the faster discharge at higher currents, such that the required dead time range of DT_{lo} does not significantly change.

6.2.4 Experimental Results

The DC-DC buck converter has been implemented in a 180 nm high-voltage BiCMOS technology. The die with an active area of around 2.5 mm^2, shown in Fig. 6.7a, was directly bonded to a PCB to minimize ringing, as shown in Sect. 3.3 (Fig. 3.6c). The PCB including the measurement setup is depicted in Fig. 6.7b.

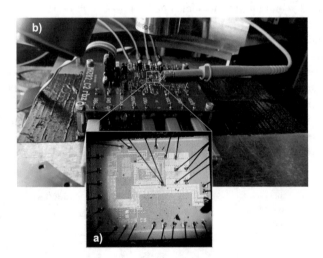

Fig. 6.7 Micro-photograph of the test chip, directly bonded to the PCB

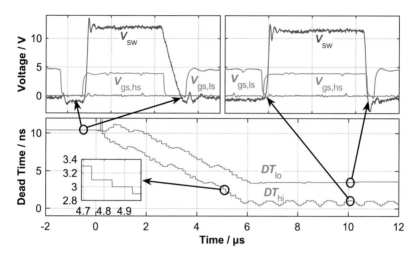

Fig. 6.8 Dead time regulation and settling over time

Initial measurements were performed with large passives $L_0 = 10\,\mu\text{H}$ and $C_0 = 10\,\mu\text{F}$ to ensure continuous conduction mode (CCM) over a wide load current range. Figure 6.8 shows the voltage at the switching node V_{sw} and the gate drive signals $V_{\text{gs,hs}}$ and $V_{\text{gs,ls}}$ before the dead time control is started (top left) and after the dead time has settled at around 6 µs (top right). From this, both dead times DT_{hi} and DT_{lo} are extracted in each period and drawn over time in the bottom graph of Fig. 6.8. Initially, the dead time is 10 ns at both transitions. After settling, the dead time DT_{hi} at turn-on of the high-side switch is around 500 ps. DT_{lo} reaches 4 ns with nearly ZVS and without low-side body diode conduction. A steady state can be recognized after 6 µs settling with a small control ripple for DT_{lo} ($\leq$2 LSB). Due to discrete time sampling of the steep rising slope at V_{sw}, DT_{hi} remains in regulation. A steady state could be achieved by widening the tunable target window width. During dead time regulation, an adjustment of DT_{hi} and DT_{lo} slightly increases the PWM duty cycle. This leads temporarily to a higher inductor current which results in a faster slope at the switching node. Due to 100 ps time resolution of the oscilloscope, a step size between 100 and 200 ps was measured (inset in Fig. 6.8, bottom), which is in the expected range of 125 ps.

Figure 6.9 shows a measurement of the entire voltage regulation at a load step from 50 to 150 mA. Before the load step (at $-0.2\,\mu\text{s}$) the dead time regulation is at its optimal value, which can be seen at V_{sw} in Fig. 6.9a. The low-side turn-on can be identified by a slight ringing at V_{sw} due to a coupling of the gate charge into the switching node. After the load step, which occurs at 0 ns, the output voltage V_{out} drops by 50 mV. Figure 6.9b shows V_{sw} at the point, where V_{out} dropped to its minimum after the load step. Due to the fast drop of V_{out}, also the inductor current increases rapidly and the falling slope of V_{sw} is significantly faster. At that point, the dead time DT_{lo} at the turn-low event is not optimal anymore as the falling slope has changed fast. Only 1.5 µs later, the dead time control has regulated again to the

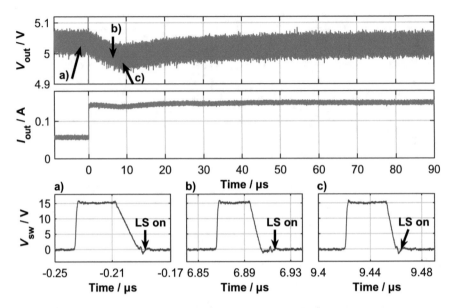

Fig. 6.9 Voltage regulation at a load step of 100 mA with the switching node signal at (**a**) before the load step, (**b**) during the voltage drop with non-optimal dead time and (**c**) with optimal dead time

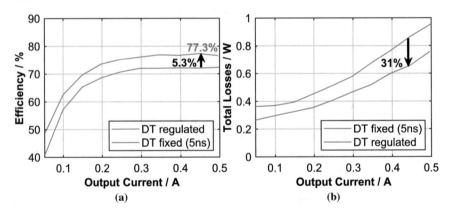

Fig. 6.10 Comparing the proposed dead time control to a fixed dead time of 5 ns: (**a**) measured efficiency of the synchronous buck converter with 18 V input and 5 V output voltage; (**b**) measured overall converter losses

optimal value, as seen in Fig. 6.9c. As the dead time control is able to adjust the dead time by 125 ps per switching period, it is fast enough to follow the slow output voltage regulation loop until V_{out} has finally settled back to its target value.

Figure 6.10 shows the efficiency and the losses of the buck converter at 18 V input, 5 V output, and 10 MHz switching for varying output currents I_{out}. A peak

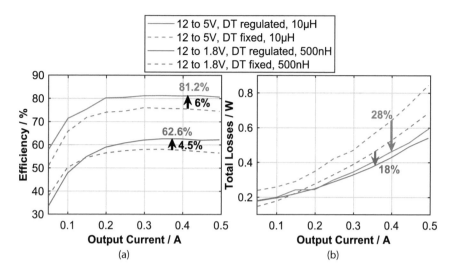

Fig. 6.11 Comparing the proposed dead time control to a fixed dead time of 5 ns: (**a**) measured efficiency of the synchronous buck converter for a 12 V to 5 V conversion with an inductor of 10 μH and a conversion from 12 V to 1.8 V with an inductor of 500 nH; (**b**) measured overall converter losses

efficiency of 77.3% is achieved. Compared to a fixed 5 ns dead time, the efficiency and the losses improve by 5.3% and 31%, respectively, due to the proposed dead time control.

A fixed dead time of 5 ns is chosen to assure no cross conduction at the power switches, which can occur due to the mismatch in the propagation delay between the low-side and high-side PWM signal path or due to temperature gradients on chip. Figure 6.11 shows the efficiency and the losses at 12 V input. At 5 V output and 10 MHz switching, the overall peak efficiency of 81% is achieved. Compared to a fixed 5 ns dead time, the efficiency improves by at least 6% (28% lower losses) over the measured load range with the dead time control. A second measurement in Fig. 6.11 is done with a conversion to 1.8 V output, in this case with an inductor of 500 nH. The efficiency dropped by nearly 20%, because the output power is less at lower output voltages, but the switching losses remain equal (refer to Sect. 5.7). Nevertheless, an improvement by the dead time control of 4.5% is achieved, which is a loss reduction of about 18%. In addition, a conversion from both 12 and 18 V input voltage to 1.8 and 5 V output voltage, respectively, was measured. The results listed in Table 6.1 do also include a smaller inductor of 500 nH.

Table 6.2 compares the results to state-of-the-art converters. The proposed converter design achieves similar or even better results in the improvement of peak efficiency at similar switching frequency and output power, but with higher input voltage range and conversion ratios. The large efficiency improvement of 6.5% in [11] is derived with respect to a large fixed dead time of 20 ns.

Table 6.1 Measurement results at different operating points

V_{in}	V_{out}	I_{out}	η_{opt}	Improvement		L_0
				$\Delta\eta$	ΔP_{losses}	
12 V	5 V	0.4 A	81.2%	6%	27.9%	10 µH
12 V	5 V	0.2 A	80.2%	6%	30.5%	10 µH
12 V	1.8 V	0.2 A	57.9%	3%	9.4%	500 nH
12 V	1.8 V	0.5 A	60.8%	5%	21%	500 nH
18 V	5 V	0.2 A	73.6%	5%	21.5%	10 µH

Table 6.2 Comparison with prior art

	Proposed converter		[5]	[11]
Switching frequency	10 MHz		10 MHz	10 MHz
Resolution	125 ps		1 ns	Analog
Efficiency improvement	6%	5%	3%	6.5%
Output current	200 mA	500 mA	120 mA	600 mA
Input voltage	12 V	18 V	5 V	3.6 V
Technology	180 nm		65 nm	350 nm

6.3 Enhanced Dead Time Control for Light Load

This section presents the implementation of a dead time control for light-load condition described in Sect. 5.4.4. The negative inductor peak current (see Fig. 5.8b) allows to achieve soft-switching on both switching transitions at high-side and low-side switch turn-on, resulting in a further switching loss reduction in a synchronous buck converter over a wide input voltage range up to 48 V and a wide load current range, even at a fixed switching frequency. The light-load dead time control utilizes the proposed dead time control implementation described in Sect. 6.2. At the falling switching node transition during DT_{lo}, the inductor current remains positive, which allows to control DT_{lo} the same way as for full-load conditions as described in Sect. 6.2. Only the implementation of DT_{hi} control needs to be adjusted, which is described subsequently.

Typically, a further reduction of the size of the main inductor in the converter results in higher current ripple, and thus in higher inductor losses (see Fig. 2.13a) and conduction losses. In contrast, a smaller inductor allows to increase the light-load range, in which soft-switching condition can be achieved at the high-side switch turn-on. The losses caused by a higher current ripple are thus compensated by the resulting lower switching losses due to soft-switching.

In light-load condition, conventional converters operate in discontinuous conduction mode (DCM) with complex loop compensation [4], or they operate in forced continuous conduction mode (FCCM) with excessive losses due to hard switching of the high-side switch. Soft-switching techniques for this case were presented, which require to operate in critical conduction mode (also known as boundary conduction mode) [3, 7, 9]. In this mode, the converters operate exactly at the boundary between DCM and CCM, i.e., the low-side switch or the freewheeling diode turns off when the inductor current is zero. The parasitic capacitances and the inductor create a voltage ringing at the switching node while the high-side switch is turned on at the maximum amplitude with reduced drain-source voltage and thus reduced switching losses. As the amplitude of the resonating switching node can achieve a maximum of twice the output voltage, soft-switching is only possible for low input voltages or requires more complex tapped inductors. Moreover, critical conduction mode can only be achieved with a variable switching frequency. In other concepts, known as Zero-Voltage Switching Quasi-Square-Wave (ZVS-QSW) converters [2, 8, 12], the inductor current is significantly forced negative by the low-side switch. When the low side is turned off, the switching node is pulled up by the negative inductor current, similarly as it is in the light-load condition described in Sect. 5.4.4. However, it is the intention to force the body diode of the high-side switch to become conducting across the entire operating range, to be able to assure that the high-side switch can be turned on with zero-voltage switching (ZVS). The disadvantage is that a very high peak current in the inductor is required, and additional losses in the forward conducting body diode of the high-side switch occur.

6.3.1 Turn-High Dead Time Control Concept

To modify the implementation of the full-load regulation of DT_{hi} to operate in light-load condition, the target window for the evaluation of the switching node voltage V_{sw} needs to be adjusted. In full-load condition, V_{sw} was sampled shortly after the low-side switch turned off (instant S_1 in Fig. 6.1), and the sampled voltage at S_1 was regulated to a target window around zero volt. To achieve ZVS or soft-switching in light-load condition, DT_{hi} has to be adjusted such that the switching node voltage V_{sw} is in a target window around V_{in}, shortly before the high-side switch turns on. If V_{sw} is above the target window around V_{in}, the body diode of the high-side switch is conducting, and thus DT_{hi} is too long, as shown in Fig. 5.11a. If V_{sw} is significantly below the target window around V_{in}, the high-side switch turns on with hard-switching, and thus the DT_{hi} is too short.

The sampling of V_{sw} used for the full-load dead time control, shown in Fig. 6.3, can be modified by changing the reference voltages of the window comparators, as well as the trigger signal for the sampling event, which is synchronous to the high-side switch turn-on. The voltage divider at V_{sw} was designed with a ratio of about 10:1 to divide the full voltage range of V_{sw} into the low-voltage domain. Thus, the target window changes from the lower range of the low-voltage domain in full load to the upper range of the low-voltage domain at light load. In case the negative current in the inductor is not strong enough to fully pull up V_{sw} to the target window around V_{in}, as shown in Fig. 5.13a ($L_0 = 500$ nH), a maximum detection circuit has to be added to turn on the high-side switch at the maximum of V_{sw} (loss optimum) with soft-switching, instead of further increasing DT_{hi} as V_{sw} is still below the target window.

A maximum detection circuit is used in the parallel-resonant converter PRC, and will be presented in Chap. 7. Its principle is shown in Fig. 7.7.

6.3.2 Experimental Results

The dead times at both switching events are adjusted by the implemented dead time generator. Measurements are performed with $L_0 = 500\,\text{nH}$ and $C_0 = 10\,\mu\text{F}$. Figure 6.12 shows the switching node voltage V_{sw} and the gate drive signals $V_{gs,hs}$ and $V_{gs,ls}$ for an optimal dead time for $I_{out} = 150\,\text{mA}$, where ZVS is achieved at both switching events. An appropriate DT_{hi} enables to charge the switching node up to V_{in} for ZVS.

A measurement of the switching node V_{sw}, depicted in Fig. 6.12, achieves ZVS at both switches with correctly adjusted dead times DT_{hi} and DT_{lo}. The switching node V_{sw} is pulled up to V_{in} after the low-side switch is turned off ($V_{gs,ls}$ switching to low), and the high-side switch is turned on with ZVS, exactly when V_{sw} reaches V_{in}. It also can be observed that ringing at V_{sw} when the high-side switch turns on is significantly improved due to ZVS, as large inrush current to charge parasitic capacitances of the switch is minimal. Figure 6.13b shows the efficiency of the proposed synchronous buck converter in dependence of the load current at 48 V input, 5 V output, and 10 MHz switching frequency. A peak efficiency of 64.3% is achieved. Figure 6.13b also shows the efficiency for the converter operating with the conventional dead time control (Sect. 6.2) (i.e., for minimized cross conduction and reverse recovery losses), with ZVS at the low-side switch and hard switching with minimum dead time at the high-side switch [10]. The efficiency improves by up to 14.4% for low output current, which relates to a loss reduction of up to 45%.

Fig. 6.12 Measured voltages at the switching node and the corresponding gate drive signals at light load with enhanced dead time control, achieving ZVS at both switching events

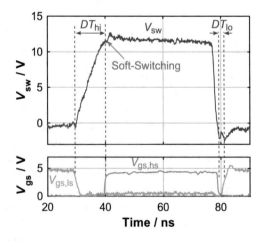

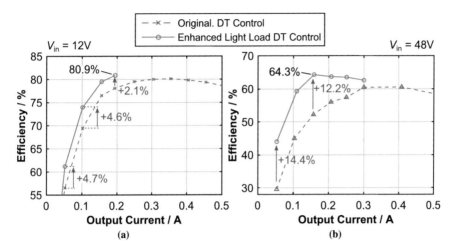

Fig. 6.13 Efficiency measurements of the enhanced light-load dead time control and comparison to the original dead time control (as proposed in Sect. 6.2 for full-load operation) over varying load current with $V_{out} = 5$ V for (**a**) $V_{in} = 12$ V, and (**b**) $V_{in} = 48$ V

For a conversion from 12 V input to 5 V output (Fig. 6.13a), soft-switching at both switches leads to a high converter peak efficiency of 80.9%, with an improvement to the conventional dead time control [10] of up to 4.7%, resulting in a loss reduction of up to 22%. The proposed dead time concept with soft-switching at both switches is most effective at high input voltages as high as 48 V.

References

1. Abu-Qahouq J, Mao H, Al-Atrash H, Batarseh I (2006) Maximum efficiency point tracking (MEPT) method and digital dead time control implementation. IEEE Trans Power Electron 21(5):1273–1281. https://doi.org/10.1109/TPEL.2006.880244
2. Chen S, Trescases O, Ng WT (2003) Fast dead-time locked loops for a high-efficiency microprocessor-load ZVS-QSW DC/DC converter. In: IEEE conference on electron devices and solid-state circuits, 2003, pp 391–394. https://doi.org/10.1109/EDSSC.2003.1283557
3. Chinag CY, Chen CL (2009) Zero-voltage-switching control for a PWM buck converter under DCM/CCM boundary. IEEE Trans Power Electron 24(9):2120–2126. https://doi.org/10.1109/TPEL.2009.2021186
4. Cuk S, Middlebmok R (1977) A general unified approach to modelling switching DC-to-DC converters in discontinuous conduction mode. In: IEEE power electronics specialists conference (PESC'77), pp 36–57. https://doi.org/10.1109/PESC.1977.7070802
5. Maderbacher G, Jackum T, Pribyl W, Wassermann M, Petschar A, Sandner C (2011) Automatic dead time optimization in a high frequency DC-DC buck converter in 65 nm CMOS. In: 2011 Proceedings of the ESSCIRC (ESSCIRC), pp 1919–1922. https://doi.org/10.1109/ESSCIRC.2011.6045013
6. Mappus S (2003) Predictive gate drive boosts synchronous DC/DC power converter efficiency. Application Report, SLUA281, Texas Instruments

7. Park JH, Cho BH (2005) The zero voltage switching (ZVS) critical conduction mode (CRM) buck converter with tapped-inductor. IEEE Trans Power Electron 20(4):762–774. https://doi.org/10.1109/TPEL.2005.850919

8. Vorperian V (1988) Quasi-square-wave converters: topologies and analysis. IEEE Trans Power Electron 3(2):183–191. https://doi.org/10.1109/63.4348

9. Wang J, Zhang F, Xie J, Zhang S, Liu S (2014) Analysis and design of high efficiency quasi-resonant buck converter. In: 2014 International electronics and application conference and exposition (PEAC), pp 1486–1489. https://doi.org/10.1109/PEAC.2014.7038084

10. Wittmann J, Barner A, Rosahl T, Wicht B (2015) A 12V 10MHz buck converter with dead time control based on a 125 ps differential delay chain. In: 41st European solid-state circuits conference (ESSCIRC), ESSCIRC 2015, pp 184–187. https://doi.org/10.1109/ESSCIRC.2015.7313859

11. Yan W, Pi C, Li W, Liu R (2010) Dynamic dead-time controller for synchronous buck DC-DC converters. Electron Lett 46(2):164–165. https://doi.org/10.1049/el.2010.2651

12. Zhou X, Wong PL, Xu P, Lee F, Huang A (2000) Investigation of candidate vrm topologies for future microprocessors. IEEE Trans Power Electron 15(6):1172–1182. https://doi.org/10.1109/63.892832

Chapter 7
Resonant Converters

Resonant converters have a long history, especially in high-power and high-voltage applications [3, 4, 7, 9–13, 15, 17–19]. In recent years, they also become common as highly integrated converter in the lower power range [22].

In resonant or quasi-resonant converters, soft-switching or zero-voltage switching is typically achieved by generating an oscillation at the switching node, caused by a resonant circuit. The resonant circuit contains an inductor, which provides an oscillating current, pulling up the switching node in a periodic manner. The turn-on of the power switches are synchronized to the oscillation in order to turn on the switch with zero-voltage or with soft-switching, as discussed in Sect. 2.2.2.4.

In resonant converters, the frequency of a resonant circuit determines the overall switching period of the converter, while in quasi-resonant converters, an oscillation of a resonant circuit only occurs during the off-time of the converter. As a consequence, efficiency is only achieved in a very narrow operating range [11, 17]. A wide range of load and input voltages would require to vary the converter's switching frequency over a wide range. Section 7.1 reveals the associated limitations for conventional resonant converters using an example of a quasi-resonant converter (QRC) architecture.

A parallel-resonant converter (PRC) architecture, including a high-resolution soft-switching control, is proposed in Sect. 7.2, which significantly reduces the required switching frequency range, and thus is able to cover a wide input voltage and load range.

7.1 Quasi-Resonant Converter

The lowest switching frequency across the operating range of a switching converter determines the size of the passive filter components if a specific limitation of the output voltage ripple needs to be fulfilled. Operating points requiring higher

© Springer Nature Switzerland AG 2020
J. Wittmann, *Integrated High-V$_{in}$ Multi-MHz Converters*,
https://doi.org/10.1007/978-3-030-25257-1_7

switching frequencies lead to additional switching losses, without further benefit in terms of converter volume. As a consequence, it is desirable to limit the switching frequency range in resonant converters. The below analysis of a conventional quasi-resonant converter [5, 14] demonstrates that boundaries of the frequency range, strongly limit the operating range (V_{in}, I_{out}). A wider operating range would only be possible at slower switching by using a significantly larger resonant circuit, which limits integration.

The architecture of a conventional quasi-resonant converter (QRC) is shown in Fig. 7.1. An IC-level design was presented in [5]. Zero-voltage switching (ZVS) can be achieved by adding a resonant circuit L_r and C_r to a regular buck converter. The converter is controlled only with one switch, which is beneficial for lower switching losses in multi-MHz operation.

The operating principle is demonstrated by means of the converter's main waveforms in Fig. 7.2. When the switch MPR is turned on, the current I_{Lr} increases through the resonant inductor L_r until it carries the output current I_{out}. The switching node V_{sw} rises as long as it is equal to the input voltage V_{in}. The diode D_0 is blocking, and the current in L_r stays constant as $V_{sw} = V_r$. When the switch MPR is turned off, V_r and V_{sw} decrease to a negative forward voltage $-V_f$ of D_0

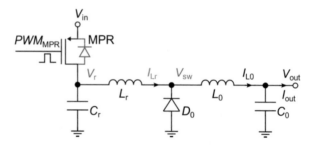

Fig. 7.1 Conventional quasi-resonant converter (QRC) [5, 14]

Fig. 7.2 Signals of the conventional quasi-resonant converter (QRC)

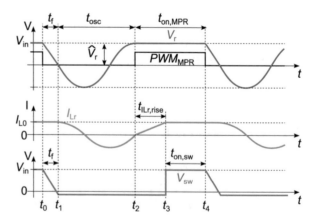

and the diode D_0 turns on. V_r begins to oscillate around $-V_f$. Once V_r reaches V_{in}, the switch MPR can be switched on with ZVS, i.e., zero drain-source voltage.

As depicted in Fig. 7.2, the overall switching period (and thus the switching frequency) $T = 1/f_{sw}$ splits up into four different periods t_f, t_{osc}, $t_{ILr,rise}$, and $t_{on,sw}$. The switching period can thus be written as

$$T = t_f + t_{osc} + t_{ILr,rise} + t_{on,sw}. \tag{7.1}$$

When the power switch MPR is turned on at t_0, V_{sw} and V_r are both discharged approximately in the same time t_f by the inductor currents I_{Lr} and I_{L0} to the value of the negative forward voltage $-V_f$ of D_0. t_f scales linearly with V_{in} and with $1/I_{out}$ (see Fig. 7.2). Thus, it significantly impacts the period and the switching frequency at varying operating points.

At time t_1, V_{sw} is held constant to $-V_f$ by the conducting diode D_0. As MPR is still off, V_r experiences an oscillation around a mean value of $-V_f$. The amplitude $\hat{V}_r$ of the oscillation calculates to

$$\hat{V}_r = \sqrt{L_r/C_r} \cdot I_{out}. \tag{7.2}$$

For $\hat{V}_r < V_{in}$, MPR can be turned on with soft-switching at the maximum oscillation of V_r, which occurs at $2/3$ of the oscillation period. In this case, t_{osc} is independent of V_{in} and I_{out}. The quality of soft-switching is related to the voltage difference across MPR when it turns on. The voltage difference depends on the amplitude $\hat{V}_r$. For $\hat{V}_r > V_{in}$, MPR turns on (or the body diode of MPR becomes conducting), as soon as the oscillation of V_r reaches V_{in}. In this case, the turn on event happens before $2/3$ of the oscillation period (before the oscillation would reach its maximum), and thus t_{osc} becomes dependent of V_{in}, and also of I_{out}, as it changes $\hat{V}_r$ (see (7.2)). At time t_2, the inductor current I_{Lr} is zero. During $t_{ILr,rise}$, I_{Lr} increases until it reaches the current in the inductor L_0. At t_3, it exceeds I_{L0} and V_{sw} is pulled up. The high-phase and thus the effective on-time $t_{on,sw}$ of the QRC ends at t_4, when MPR is turned off again. $t_{on,sw}$ depends on V_{in}, as it defines the duty cycle of the converter.

The impact of the operating points on all the converter phases results in a wide variation of both the switching frequency, as well as the effective on-time $t_{on,sw}$ of the converter. In Fig. 7.3, the frequency variation of the conventional QRC is plotted over a wide input voltage and load current range, while the resonant components are $L_r = 300$ nH and $C_r = 30$ pF, which are in the range to be potentially fully integrated. A variation of V_{in} from 10 to 50 V, and of the load from 0.1 to 0.5 A results in a five times frequency variation. Resonant converters were published, which use an external tunable capacitor to adjust the resonant frequency [6].

One limitation of the QRC is that V_r swings negative during t_{osc} to $-\hat{V}_r$. The drain-source voltage of MPR thus reaches a maximum of $V_{in} + \hat{V}_r$. To achieve zero-voltage switching, requiring $\hat{V}_r > V_{in}$, the maximum ratings of MPR need to be $>2V_{in}$. Assuming that a realistic design for a 50 V input will use a technology rated to 100 V, the maximum amplitude has to be limited to $\hat{V}_r = 50$ V. This

Fig. 7.3 Frequency variation in a frequency-controlled quasi-resonant converter for a wide load and input voltage range

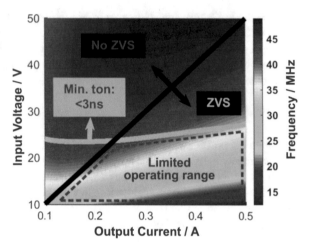

corresponds to the maximum load current of $I_{out} = 0.5$ A for the design in Fig. 7.3. The maximum $\hat{V}_r$ decreases towards smaller loads, and ZVS is achieved only for a combination of small input voltage and large output load. This is indicated in Fig. 7.3 (diagonal line).

A second limitation is related to the fact that the required duty cycle becomes very small at high V_{in}. Since high V_{in} requires high switching frequencies at the same time, the resulting effective on-time $t_{on,sw}$ is reduced to the sub-nanosecond range. The operating range, in which the on-time reduces to below 3 ns, is indicated in Fig. 7.3. In high-voltage technologies, which are typically not optimized for speed, the generation of on-time pulses <3 ns is usually very challenging, as demonstrated in Chap. 4. Together with the ZVS limitation, the QRC can only support a significantly reduced operating range, which is indicated in Fig. 7.3.

7.2 Parallel-Resonant Converter (PRC)

A parallel-resonant converter is proposed in this section, which overcomes the limitations of conventional resonant converters. The proposed parallel-resonant converter consists of a conventional buck converter with MN0, external diode D_0, and output filter L_0, C_0, controlled by PWM_{MN0}. D_0 is implemented as a Schottky diode to avoid reverse recovery losses and to reduce forward conduction losses. The output voltage is controlled by the on-time of MN0, identical to a conventional buck converter. Thus, the output voltage regulation is state-of-the-art.

7.2.1 Switching Concept and Operating Modes

Resonant operation is achieved by placing a resonant circuit L_r, C_r, and a second switch MPR in parallel to a buck converter (Fig. 7.4). A PMOS switch is used as the resonant node reaches negative voltage levels of $V_r = -V_{in}$, which is not supported by NMOS devices in various technologies. The resonant switch MPR controls the energy in the resonant circuit. Figure 7.5 (Mode 1) shows the signals of the switching node V_{sw} of the buck converter stage and V_r of the resonant circuit. MPR is turned on at t_0. As V_{sw} is still low, the current in the inductor L_r increases until it exceeds the current through L_0 and starts to charge up V_{sw} towards t_2 with nearly no losses.

MN0 is turned on with ZVS, when V_{sw} reaches V_{in}, and the turn-on losses of MN0 are significantly reduced. The maximum voltage of V_{sw}, and thus the minimum drain-source voltage $V_{ds,MN0}$ at turn-on of MN0, is determined by the on-time of MPR. Thus, the on-time of MPR controls ZVS of MN0. After MPR is turned off (t_1), V_r resonates and swings back towards V_{in} after one or multiple oscillation periods n, which allows to also turn on MPR with soft-switching at t_4. The amplitude of the resonant circuit depends on V_{in}, I_{out}, and C_r. To achieve soft-switching at MPR over a wide operating range (V_{out} and I_{out}), C_r is adjustable, implemented as an integrated 5-bit capacitor array.

Smaller C_r values lead to higher amplitudes, required for increasing V_{in}, which results in a higher resonant frequency. At the same time, higher V_{in} requires smaller duty cycles, which is limited by the resonant charging of V_{sw}. To overcome this limitation, the converter operates with different numbers of oscillation periods n of V_r, depending on the value of I_{out} and V_{in} (and thus C_r). Three different operation modes can be distinguished in the PRC, depending on the turn-on conditions of MPR and the number of oscillation periods n of V_r. The signals of the different operation modes are shown in Fig. 7.5.

Mode 1: The turn-on of MPR occurs after a single oscillation of V_r ($n = 1$). C_r adjusts the maximum of V_r to reach V_{in} at the turn-on of MPR. V_{in} is limited to <24 V, as the on-time of MN0 becomes small (<3 ns). ZVS brings good efficiency, despite the high frequency with considerable gate charge losses.

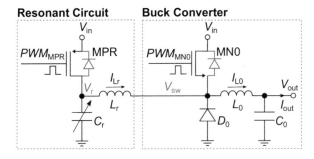

Fig. 7.4 Architecture of the proposed parallel-resonant converter (PRC) [24, 25]

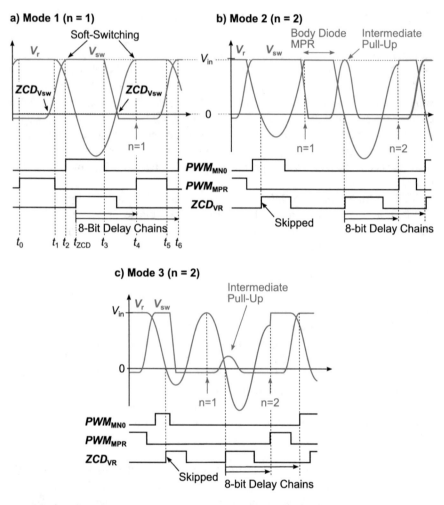

Fig. 7.5 Signals and switching principle of the proposed parallel-resonant converter (PRC) in its different operation modes 1–3 (**a–c**)

Mode 2: C_r is set to its maximum value to achieve a lower oscillation frequency of V_r, resulting in reduced switching losses. The turn-on of MPR happens after two oscillation periods of V_r ($n = 2$). The entire first period occurs while MN0 is on. Consequently, V_r is oscillating around V_{in} and the positive half-wave (after $n = 1$) is clamped by the body diode of MPR. As V_{sw} is low, the resonance inductor L_r experiences a significant current increase during body diode conduction and the resulting high-phase of V_r. The current in the resonant inductor I_{Lr} is high enough to again pull up V_{sw}. This intermediate swing up of V_{sw} again transfers energy to the output without turning on MN0, which reduces the overall voltage ripple at the output. A conventional buck converter requires to switch up to 40% faster to

Table 7.1 Operation modes of the PRC

	Oscillation cycles n	V_{in} (typ.)	Switching frequency (typ.)
Mode 1	1	<24 V	15–30 MHz
Mode 2	2	<25 V	9–15 MHz
Mode 3	2	25–36 V	11–14 MHz
	≥ 3	>36 V	9–10 MHz

achieve the same output voltage ripple at $n = 2$. V_r swings up close to V_{in}, and soft-switching of MPR is achieved.

Mode 3: The value of C_r is set low enough (higher frequency) such that V_r reaches V_{in} at $n = 1$ and $n = 2$. With higher switching frequency and lower intermediate V_{sw} pull-up, the output voltage ripple is similar to Mode 2. A frequency variation of +/-20% changes the ripple by only +/-8%. $V_{in} = 48$ V can be handled for $n = 3$ at reduced frequency and losses.

Table 7.1 summarizes the operation modes, including the selection of n, the related V_{in} range for each of the modes and the according switching frequency range.

7.2.2 Converter Implementation

Figure 7.6 shows the implemented soft-switching control for MPR and MN0. When the falling edge of V_r crosses zero, a zero-cross detection circuit sets ZCD_{Vr} high. In each switch control block (bottom part of Fig. 7.6) the positive edge of ZCD_{Vr} is passed through two variable 8-bit delay chains with 250 ps resolution (LSB), which results in a maximum possible delay of 64 ns. The delayed signal ($turn\_on$) initiates PWM_0 or PWM_r to be set high. The delay is adjusted cycle-by-cycle by an up-down counter, such that both MN0 and MPR turn on with minimum voltage across (soft-switching or ZVS).

To evaluate if soft-switching is achieved, V_{sw} and V_r are each sampled twice, with a delay of about 500 ps, by the maximum detection block (Fig. 7.7), shortly before the corresponding switches are turned on. If the first sampled voltage S_1 is smaller than the second sampled voltage S_2, MN0 or MPR have been turned on too early at the rising edge of V_{sw} or V_r, before the maximum voltage is achieved. Consequently, the up-down counter, controlling the 8-bit delay, is increased for the next period. If $S_1 > S_2$, MN0 and MPR have been turned on too late, after the maximum of V_{sw} or V_r, the counter is decreased. The counter is also decreased if $S_1 \approx S_2$. This case indicates either that the body diode is conducting (turn-on too late), or the corresponding switch is turned on at the optimum soft-switching point, which is the maximum of V_r or V_{sw}. Consequently, the 8-bit delay value is toggled between the optimum turn-on point and one LSB time step after the optimum turn-on point. A logic-based range monitor observes ZCD_{Vsw} and ZCD_{Vr} to ensure that the turn-on only occurs if V_{sw} and V_r are positive.

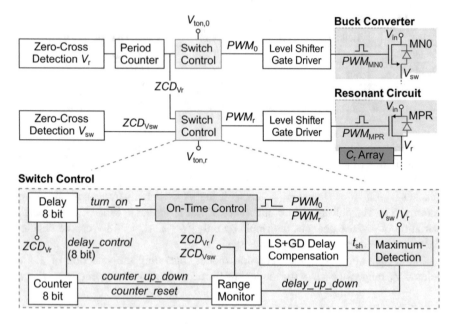

Fig. 7.6 Block diagram of the proposed parallel-resonant converter

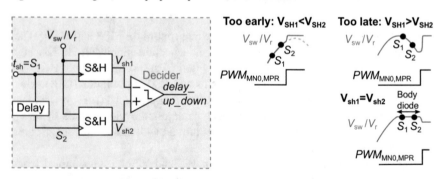

Fig. 7.7 Maximum detection circuit and principle

The on-time control (Fig. 7.8) operates similar to a conventional PWM generator. The signal *turn_on* initiates an integrator ramp. The on-times of MN0 or MPR are controlled by comparing the ramp to the reference voltages $V_{ton,0}$ or $V_{ton,r}$. $V_{ton,0}$ controls the output voltage, while $V_{ton,r}$ is adjusted such that MP0 is turned on with ZVS. In the implementation of the PRC, the proposed PWM generator (see Sect. 4.2) is re-used, its logic is re-configured to use one PWM generator stage for the generation of PWM_0, and the second stage for generation of PWM_r.

The integrated 5-bit capacitor array, which covers a range of 50–150 pF, is shown in Fig. 7.9. Its LSB element utilizes interleaved finger capacitors with a poly-Si plate against substrate coupling and associated losses. If the high-voltage switch

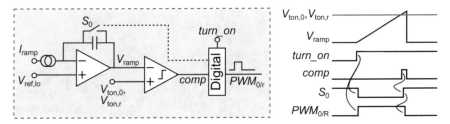

Fig. 7.8 On-time control of PWM_0 and PWM_r, circuit and signals

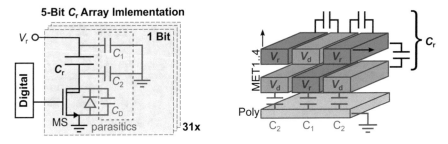

Fig. 7.9 Implementation of the 5-bit C_r array

MS is turned on, the capacitor cell is enabled and increases the total C_r, resulting in a lower switching frequency and higher amplitudes of V_r, adjusting the soft-switching condition. Parasitic capacitors determine the minimum value of C_r (if all MS switches are deselected).

The sample and hold (S&H) circuit of the maximum detection blocks is re-used from the dead time implementation of Chap. 6, Fig. 6.3.

Figure 7.10 shows the zero-cross detection (ZCD) circuits for both the resonant node V_r (Fig. 7.10a) and the switching node V_{sw} (Fig. 7.10b). Each ZCD circuit comprises an additional frequency compensated voltage divider, similar to the divider in the S&H stage. V_r requires a more accurate zero-cross detection for the period counter, while V_{sw} uses a simple inverter with clamped input for high-voltage protection, which connects to the range monitor.

7.2.3   Experimental Results

The proposed PRC has been implemented in a 180 nm high-voltage BiCMOS technology (Fig. 7.11a). Figure 7.11b shows the PCB of the converter used for the experiments.

Figure 7.12a shows the measured transient signals at $V_{in} = 16$ V in Mode 2 ($n = 2$, see Fig. 7.5) with the body diode conducting at the first maximum of the oscillation of V_r and the resulting high intermediate pull-up close to V_{in} reducing the

Fig. 7.10 Zero-cross detection circuits for (**a**) V_r, and (**b**) V_{sw}

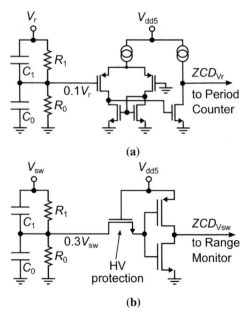

(**a**)

(**b**)

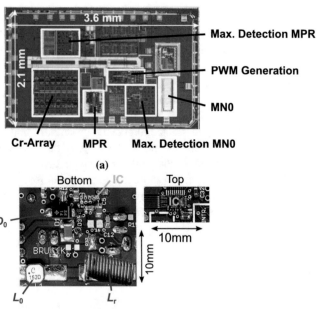

(**a**)

(**b**)

Fig. 7.11 Photograph of (**a**) the test chip implemented in a 180 nm high-voltage BiCMOS technology, and (**b**) the board of the converter with the IC directly bonded to the PCB

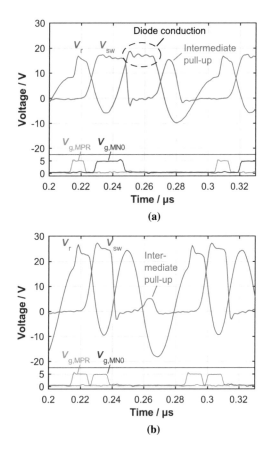

Fig. 7.12 Transient measurements of the proposed parallel-resonant converter. (**a**) $V_{in} = 16$ V, $I_{out} = 300$ mA, Mode 2 with $n = 2$. (**b**) $V_{in} = 24$ V, $I_{out} = 300$ mA, Mode 3 with $n = 2$

output voltage ripple. In Fig. 7.12b, the converter operates at $V_{in} = 24$ V in Mode 3 ($n = 2$), without body diode conduction at a higher overall switching frequency and a reduced intermediate pull-up.

For disabled soft-switching control, Fig. 7.13a confirms inefficient hard-switching of both devices at the switching node V_{sw} and the resonant node V_r. Figure 7.13c depicts the measured settling of the turn-on delay for MN0 and MPR, respectively, after the soft-switching control is enabled. Both delays increase and reach their optimum values (soft-switching) at 87 μs ($t_{d,sw}$) and 73 μs ($t_{d,r}$). The according signals V_{sw} and V_r in soft-switching condition are shown in Fig. 7.13b for $V_{in} = 15$ V, 300 mA and $n = 2$ oscillations.

Figure 7.14 depicts the measured efficiency vs. V_{in}. At $I_{out} = 500$ mA (Fig. 7.14a), the peak efficiency of the PRC is 76.3% at $V_{in} = 24$ V in Mode 2 at a switching frequency of 10.8 MHz, and 71.5% in Mode 1 at a switching frequency close to 25 MHz. Thus, Mode 1 is slightly less efficient but allows up to a 2.5× increase of the switching frequency, resulting in a significantly smaller output filter. However, Mode 1 is limited to a maximum input voltage of 24 V at $I_{out} = 500$ mA and 18 V at $I_{out} = 300$ mA (Fig. 7.14b). Input voltages up to 48 V are reached for $n = 3$ cycles at $V_{out} = 5$ V in Mode 3 (in accordance to Table 7.1). At this

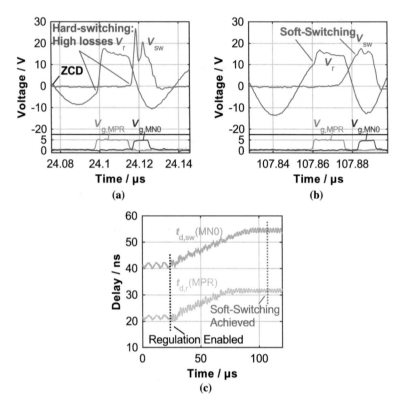

Fig. 7.13 Transient measurement results of the turn-on regulation and waveforms: (**a**) Soft-switching control disabled resulting in high switching losses; (**b**) settled soft-switching control at both MPR and MN0; (**c**) settling of the turn-on delay control for MPR and MN0

large conversion ratio of nearly 10:1, the converter still achieves efficiencies above 67%. For $V_{in}>20\,V$, the PRC is superior in efficiency by up to 10.7% compared to a conventional buck converter at a switching frequency that results in the same output voltage ripple. At $I_{out}=300\,mA$ (Fig. 7.14b) the maximum efficiency gain is 14.5% at $V_{in}=36\,V$. The two plots in Fig. 7.14 reveal that the switching frequency increases towards higher V_{in} up to 40 V to about 14 MHz for $n=2$. As soon as the number of oscillation periods of V_r is increased to $n=3$ at $V_{in}=40\,V$ and above, the switching frequency is reduced to below 10 MHz, similar to switching frequencies required at low V_{in}. This way, the overall switching frequency stays in a range of 9–15MHz, which is significantly lower than the frequency range of a conventional quasi-resonant converter, as indicated in Fig. 7.3.

Figure 7.15 shows the efficiency vs. output current. At $V_{in}=24\,V$, the proposed PRC shows up to 11.3% higher efficiency than a hard-switching buck converter, corresponding to a 21.3% loss reduction. The switching frequency increases towards lower currents for $n=2$ to nearly 15 MHz. If the number of oscillation periods is

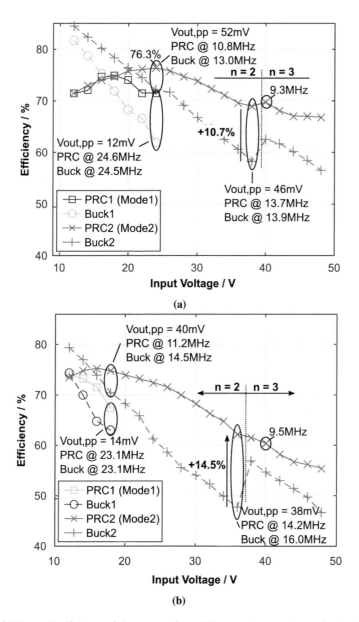

Fig. 7.14 Measured efficiency of the proposed parallel-resonant converter vs. input voltage at (**a**) $I_{\text{out}} = 500\,\text{mA}$ and (**b**) $I_{\text{out}} = 300\,\text{mA}$

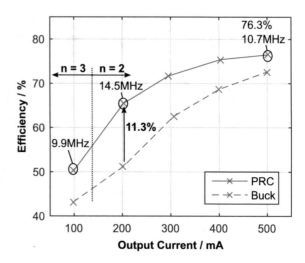

Fig. 7.15 Measured efficiency of the proposed parallel-resonant converter vs. load current at $V_{in} = 24\,V$

Table 7.2 Loss comparison of a buck converter and the PRC at $I_{out} = 0.5\,A$

	$V_{in} = 24\,V$	$V_{in} = 48\,V$
Losses in buck converter	969 mW	1918 mW
− Loss reduction by soft-switching of MN0	−586 mW	−1540 mW
+ Losses of resonant circuit	+392 mW	+862 mW
Losses of PRC	776 mW	1240 mW
Δ Losses PRC vs. buck converter	−193 mW	−678 mW
	−20%	−35%

changed to $n = 3$ at $I_{out} < 200\,mA$, the switching frequency is again reduced to about 10 MHz, which limits the switching losses at light load.

Table 7.2 shows the improvement of the losses in the PRC compared to the buck converter. It opposes the losses, which are reduced by achieving soft-switching in the buck converter stage, and the additional losses dissipated in the required PRC resonant circuit. Soft-switching nearly eliminates transition losses and charging losses of the parasitic capacitances at the switching node V_{sw}. The added losses caused by the resonant circuit are mainly due to the resonant switch. Negative voltages at V_r require the implementation of MPR as a PMOS device in this technology, with large parasitics (gate charge, etc.) and on-state resistance. Replacing MPR with an NMOS switch (by choosing a technology which supports negative voltages at the NMOS source, e.g., SoI technology), calculations indicate that the PRC has the potential to achieve peak efficiencies of >85%. The loss contribution of the integrated capacitor is negligible (typically ≤1% efficiency) due to the low-resistive implementation (see Fig. 7.9). The prototype for the experimental measurements uses an air-core inductor L_r in the resonant circuit, with inductor losses at a non-significant level. A typical resonant frequency of L_r above 20 MHz brings L_r to the border, at which air-core inductors become favorable, as shown in Sect. 2.3.3 and [16]. Further volume reduction or even on-chip integration is expected to be possible with a custom inductor design and, in general, with the ongoing progress regarding RF inductor technology and core materials. A further loss optimization is possible by replacing the external Schottky diode by an integrated low-side switch. The resulting synchronous output stage is expected to be more efficient only with an accurate dead time control [23], which makes the overall control of the converter complex.

Table 7.3 shows the comparison to prior art. Funk et al. [5] comprises a quasi-resonant concept, but soft-switching is achieved only at a fixed conversion ratio. The converter in [26] can handle 100 V, but it runs at a higher output voltage and a much lower frequency, whereby the current ripple in the inductor increases up to 2 A. To achieve similar inductor losses, this results in a multiple times larger inductor volume compared to the PRC proposed in this book, even at the same inductor value of $L_0 = 1.5\,\mu H$ (as demonstrated in Sect. 2.3.3).

Moreover, a larger output capacitor C_0 is required ($C_0 = 4.7\,\mu F$). Thanks to the resonant operation at high switching frequency with significantly lower current ripple, the PRC allows to reduce C_0 to tiny values ($C_0 = 50\,nF$), which could be easily integrated on-chip with a suitable technology, e.g., deep trench capacitors [2].

The presented PRC supports a wide input range of 12–48 V and 100–500 mA output current range at switching frequencies up to 15 MHz. Up to input voltages of 24 V, even switching frequencies up to 25 MHz with only slightly lower efficiencies are possible by changing the operation mode.

Figure 7.16 shows a comparison of the PRC and the proposed synchronous converters from Chap. 6 to state-of-the-art converters with integrated power switches, by using the design indicator proposed in Sect. 5.7. The performance of the buck converters is benchmarked along the efficiency contour lines (grey lines). Converters located at the same efficiency contour line indicate a similar efficiency performance as described in Sect. 5.7. Higher efficiency contour lines indicate that one converter would have a higher efficiency if the operating point is chosen the same (same design indicator value) as the compared converter.

Table 7.3 Comparison with prior art

	[26]	[23]	[5]	[20]	This book
Technology	0.5 μm 120V CMOS	0.18 μm HV BCD	0.18 μm HV BCD	Not reported	0.18 μm HV BCD
f_{sw}	2 MHz	10 MHz	8 MHz	5 MHz	9–15 MHz
V_{in}	12–100 V	12–18 V	20 V	8–14 V	12–48 V
I_{out}	0.5 A	0.5 A	0.2 A	10 A	0.5 A
V_{out}	10 V	1.8 V/5 V	5 V	1.2 V	5 V
L_0	1.5 μH	10 μH	2.2 μH	2×100 nH	1.5 μH
C_0	4.7 μF	10 μF	30 μF	91 nF	50 nF
Add. passives	$C = 1\,\mu F$	–	$L = 1.2\,\mu H$	$C = 1\,\mu F$	$L = 300\,nF$
Peak eff.	90%	81%	72.3%	82.5%	76.3%

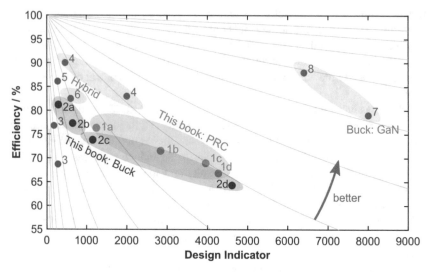

Fig. 7.16 Comparison of the converters of this book to state-of-the-art converters using the proposed design indicator

		V_{in}/V	V_{out}/V	I_{out}/A	f_{sw}/MHz	Remarks
1a	This book: PRC	24	5	0.5	10.8	
1b	This book: PRC	24	5	0.5	24.6	
1c	This book: PRC	38	5	0.5	13.7	
1d	This book: PRC	48	5	0.5	9.26	
2a	This book: Buck	12	5	0.4	10	sync. buck
2b	This book: Buck	18	5	0.45	10	sync. buck
2c	This book: Buck	24	5	0.05	10	sync. buck, light-load
2d	This book: Buck	48	5	0.15	10	sync. buck, light-load
3	Y.Ahn TPE12 [1]	3	2	0.3	50	low V_{in}, high f_{sw}
4	J.Xue JSSC16 [26, 27]	100	10	0.4	2	3-level buck converter
4	J.Xue JSSC16 [26, 27]	48	10	0.4	2	3-level buck converter
5	M.K.Song ISSCC14 [21]	3.3	1.6	0.4	40	low V_{in}
6	P.S.Shenoy APEC16 [20]	12	1.2	5	5	Remarks
7	Y.Zhang TPE16 [28]	20	5	0.8	100	buck with GaN switches
8	X.Ke ISSCC16 [8]	40	5	0.4	20	buck with GaN switches

Several operating points from the PRC are depicted (data points 1a–d in Fig. 7.16) and compared to relevant high-V_{in} multi-MHz converters. The efficiency of the PRC at operating points 1c and 1d achieves the highest design indicator value. Following the efficiency contour line towards converter 4, it indicates that the proposed PRC achieves a similar efficiency performance as 4, but at a significantly higher switching frequency with respect to its high input voltage of up to 48 V, enabling a significantly smaller size of the passives as 4. Converter 2 shows operating points of the proposed synchronous buck converter with dead time control (see Sect. 6.2) with 2a and 2b measured at low input voltages of 12 and 18 V at lower contour lines, while 2d is measured at 48 V appearing at a significantly higher contour line. The increase towards higher contour lines with higher input voltages indicates that the proposed dead time control becomes especially effective at higher input voltages. 2d is operated in light-load condition using the proposed enhanced dead time control (see Sect. 6.3). Soft-switching at both the high-side and low-side switch turn-on brings the converter to a comparable performance as the PRC, but limited to light-load range, while the PRC is able to cover the full load range.

Converter 4 and 6 are hybrid converters as described in Sect. 2.2.2.5. The comparison shows that hybrid converters achieve a comparable efficiency performances as the converters covered in this book. However, an operation at both high switching frequencies and high input voltages according to the target of this book was not demonstrated. It can be concluded that an application of the switching techniques and fast-switching circuit block proposed in this book, potentially could enable hybrid converter to achieve a similar performance at comparable design indicator values as the converters proposed in this book.

Finally, two further operating points are depicted in Fig. 7.16, which relate to published converters using external gallium-nitride (GaN) power switches. It shows that their significantly lower switching losses, as indicated in the analysis of switch technologies in Sect. 3.2, outperform converters with integrated silicon switches in efficiency. GaN switches have the potential to overcome the efficiency limitation of integrated silicon switches, and to push the boundary to significantly higher switching frequencies and conversion ratios. This demonstrates the importance of fast-switching circuit design, as covered in this book, for even higher switching frequencies and input voltages.

References

1. Ahn Y, Nam H, Roh J (2012) A 50-MHz fully integrated low-swing buck converter using packaging inductors. IEEE Trans Power Electron 27(10):4347–4356. https://doi.org/10.1109/TPEL.2012.2192136
2. Aminulloh A, Kumar V, Yang SM, Sheu G (2013) Novel structure of deep trench capacitor with higher breakdown and higher capacitance density for low dropout voltage regulator. In: 2013 IEEE 10th international conference on power electronics and drive systems (PEDS), pp 389–392. https://doi.org/10.1109/PEDS.2013.6527050
3. Chen G, Deng Y, He X, Wang Y, Zhang J (2016) Zero-voltage-switching buck converter with low-voltage stress using coupled inductor. IET Power Electron 9(4):719–727. https://doi.org/10.1049/iet-pel.2015.0267
4. Dallago E, Quaglino R, Sassone G (1996) Single-cycle quasi-resonant converter with controlled timing of the power switches. IEEE Trans Power Electron 11(2):292–298. https://doi.org/10.1109/63.486178
5. Funk T, Wittmann J, Rosahl T, Wicht B (2015) A 20V, 8MHz resonant DCDC converter with predictive control for 1ns resolution soft-switching. In: 2015 IEEE international symposium on circuits and systems (ISCAS), pp 1742–1745. https://doi.org/10.1109/ISCAS.2015.7168990
6. Guo B, Dwari S, Yongduk L, Mantese J, McCabe B, Ritter A, Nies C, Priya S, Ngo K, Zhang L, Burgos R (2017) Resonant filter based buck converters with tunable capacitor. In: 2017 IEEE energy conversion congress and exposition (ECCE), pp 2036–2042. https://doi.org/10.1109/ECCE.2017.8096407

7. Hua G, Lee FC (1995) Soft-switching techniques in PWM converters. IEEE Trans Ind Electron 42(6):595–603. https://doi.org/10.1109/41.475500
8. Ke X, Sankman J, Song MK, Forghani P, Ma DB (2016) A 3-to-40 V 10-to-30 MHz automotive-use gan driver with active BST balancing and VSW dual-edge dead-time modulation achieving 8.3% efficiency improvement and 3.4 ns constant propagation delay. In: 2016 IEEE international solid-state circuits conference (ISSCC), pp 302–304. https://doi.org/10.1109/ISSCC.2016.7418027
9. Lee FC (1988) High-frequency quasi-resonant converter technologies. Proc IEEE 76(4):377–390. https://doi.org/10.1109/5.4424
10. Lee DY, Lee BK, Yoo SB, Hyun DS (2000) An improved full-bridge zero-voltage-transition PWM DC/DC converter with zero-voltage/zero-current switching of the auxiliary switches. IEEE Trans Ind Appl 36(2):558–566. https://doi.org/10.1109/28.833774
11. Liu KH, Lee FCY (1990) Zero-voltage switching technique in DC/DC converters. IEEE Trans Power Electron 5(3):293–304. https://doi.org/10.1109/63.56520
12. Maksimovic D, Cuk S (1991) Constant-frequency control of quasi-resonant converters. IEEE Trans Power Electron 6(1):141–150. https://doi.org/10.1109/63.65012
13. Maksimovic D, Cuk S (1991) A general approach to synthesis and analysis of quasi-resonant converters. IEEE Trans Power Electron 6(1):127–140. https://doi.org/10.1109/63.65011
14. Mousavian H, Bakhshai A, Jain P (2016) An improved PDM control method for a high frequency quasi-resonat converter. 2016 IEEE energy conversion congress and exposition (ECCE), pp 1–8. https://doi.org/10.1109/ECCE.2016.7854846. http://ieeexplore.ieee.org/document/7854846/
15. Park J, Kim M, Choi S (2014) Zero-current switching series loaded resonant converter insensitive to resonant component tolerance for battery charger. IET Power Electron 7(10):2517–2524. https://doi.org/10.1049/iet-pel.2013.0757
16. Perreault DJ, Hu J, Rivas JM, Han Y, Leitermann O, Pilawa-Podgurski RCN, Sagneri A, Sullivan CR (2009) Opportunities and challenges in very high frequency power conversion. 2009 Twenty-fourth annual IEEE applied power electronics conference and exposition, pp 1–14. https://doi.org/10.1109/APEC.2009.4802625. http://ieeexplore.ieee.org/document/4802625/
17. Rashid MH (ed) (2011) Power electronics handbook: devices, circuits, and applications, 3rd edn. Elsevier Butterworth-Heinemann, Oxford
18. Ryan MJ, Brumsickle WE, Divan DM, Lorenz RD (1998) A new ZVS LCL-resonant push-pull DC-DC converter topology. IEEE Trans Ind Appl 34(5):1164–1174. https://doi.org/10.1109/28.720458
19. Sharifi S, Jabbari M (2014) Family of single-switch quasi-resonant converters with reduced inductor size. IET Power Electron 7(10):2544–2554. https://doi.org/10.1049/iet-pel.2013.0615
20. Shenoy PS, Lazaro O, Ramani R, Amaro M, Wiktor W, Khayat J, Lynch B (2016) A 5MHz, 12V, 10A, monolithically integrated two-phase series capacitor buck converter. In: 2016 IEEE applied power electronics conference and exposition (APEC), pp 66–72. https://doi.org/10.1109/APEC.2016.7467853
21. Song MK, Sankman J, Ma D (2014) A 6 A 40 MHz four-phase ZDS hysteretic DC-DC converter with 118 mV droop and 230 ns response time for a 5 A/5 ns load transient. In: 2014 IEEE international solid-state circuits conference digest of technical papers (ISSCC), pp 80–81. https://doi.org/10.1109/ISSCC.2014.6757346
22. Wei K, Ma DB (2017) Comparative topology and power loss study for high power density and high conversion ratio integrated switching power converters. In: 2017 IEEE 8th Latin American symposium on circuits systems (LASCAS), pp 1–4. https://doi.org/10.1109/LASCAS.2017.7948055
23. Wittmann J, Barner A, Rosahl T, Wicht B (2016) An 18 V input 10 MHz buck converter with 125 ps mixed-signal dead time control. IEEE J Solid State Circuits 51(7):1705–1715. https://doi.org/10.1109/JSSC.2016.2550498
24. Wittmann J, Funk T, Rosahl T, Wicht B (2017) A 12-48 V wide-Vin 9-15 MHz soft-switching controlled resonant DCDC converter. In: ESSCIRC 2017-43rd IEEE European solid state circuits conference, pp 348–351. https://doi.org/10.1109/ESSCIRC.2017.8094597

25. Wittmann J, Funk T, Rosahl T, Wicht B (2018) A 48-V wide-V_{in} 9-25-MHz resonant DC-DC converter. IEEE J Solid State Circuits 53(7):1936–1944. https://doi.org/10.1109/JSSC.2018. 2827953
26. Xue J, Lee H (2016) A 2 MHz 12-to-100 V 90%-efficiency self-balancing ZVS three-level DC-DC regulator with constant-frequency AOT V2 control and 5 ns ZVS turn-on delay. In: 2016 IEEE international solid-state circuits conference (ISSCC), pp 226–227. https://doi.org/ 10.1109/ISSCC.2016.7417989
27. Xue J, Lee H (2016) A 2 MHz 12–100 V 90% efficiency self-balancing zvs reconfigurable three-level DC-DC regulator with constant-frequency adaptive-on-time V^2 control and nanosecond-scale ZVS turn-on delay. IEEE J Solid State Circuits 51(12):2854–2866. https:// doi.org/10.1109/JSSC.2016.2606581
28. Zhang Y, Rodríguez M, Maksimović D (2016) Very high frequency PWM buck converters using monolithic GaN half-bridge power stages with integrated gate drivers. IEEE Trans Power Electron 31(11):7926–7942. https://doi.org/10.1109/TPEL.2015.2513058

Chapter 8
Conclusion and Outlook

Section 8.1 summarizes and concludes the content and results of this book. All proposed and implemented converters of this book are compared to the state-of-the-art in Sect. 8.2. In Sect. 8.3, an outlook describes possibilities and requirements on how high-V_{in} multi-MHz converters can be further improved.

8.1 Conclusion

A trend towards higher system level supply voltages is observed in several applications and systems like automotive, IT servers, automation and industrial, and others. This is mainly driven by an increasing power demand and new functionalities. Examples are the introduction of the automotive 48 V board net, in addition the 12 V battery supply, to enable high-power applications, or the increase of the server supply from 12 to 48 V to reduce wiring costs in server racks. This results in a significantly increasing input voltage for point-of-load DC-DC converters, which supply subsequent microchips and electronic systems.

Highly integrated inductive converters are beneficial to cover the requirements of high system supply voltages. Inductive buck converter fundamentals and an empirical study of the converter passives demonstrated that the switching frequency and current ripple, as well as the output voltage and the conversion ratio, have a main influence on the converter size and efficiency. At increasing switching frequencies up to 30 MHz, the switching converter passives, especially the inductors, benefit not only from a smaller size and price, but also from lower inductor losses if the current ripple is designed to be in an optimal region. A study of commercially available and published state-of-the-art converters showed that converters are rarely available for input voltage towards 50 V and switching frequencies >5 MHz at the same time. This is due to limitations of fast-switching high-V_{in} converters, which are mainly set by the switching losses, and significantly reduced minimum PWM on-time required

© Springer Nature Switzerland AG 2020
J. Wittmann, *Integrated High-V_{in} Multi-MHz Converters*,
https://doi.org/10.1007/978-3-030-25257-1_8

to control the power switch at low duty cycles (high conversion ratios). Minimum on-time pulses are limited by the circuit design, as well as by a technology related limitation of the switching node voltage slopes. This results in a structural limitation of the switching frequency towards 30 MHz for input voltages in the range of 50 V.

A preference for an asynchronous or a synchronous buck converter can be hardly distinguished as the achieved efficiency depends on the operating point, on the available switch technologies, and on the implementation of the high-side supply generation. A synchronous converter is superior in efficiency only if the dead time is precisely regulated across varying operating points. Fully integrated NMOS switches have at least five times higher switching losses compared to off-chip power switches in silicon and gallium-nitride (GaN) technologies. However, integrated switches benefit from faster and more robust switching transitions thanks to lower parasitics, especially with an appropriate assembly technology. Conventional IC packages can cause a destructive ringing, requiring a reduction of the switching slopes, which results in a limitation of the switching frequency or of the conversion ratio. Direct-bond to PCB assembly is one option to reduce ringing at the high-side gate driver supply to approximately 20%. NMOS switches are preferred due to lower switching losses, compared to PMOS switches. However, NMOS high-side switches require a floating supply, referenced to the switching node. Fast switching transitions cause an interference due to charge coupling into the signal path of the level shifter and into the substrate. Substrate coupling and isolation structures under real high-voltage switching conditions of a buck converter were investigated with both TCAD simulations and experimental measurements. Among various substrate isolation and deviation structures, back-side metalization and conducting deep-trenches are most suitable and even mandatory to effectively eliminate the impact of substrate coupling on other circuits.

The proposed designs of this book for fast-switching circuit blocks allow to control the buck converter with minimum on-time pulses of less than 3 ns, which enables an operation at input voltages up to 50 V at output voltages below 5 V and switching frequencies up to 30 MHz. A digitally multiplexed PWM generator with interleaved sawtooth signals significantly reduces the minimum possible on-time pulse. Experimental results demonstrated a generation of on-time pulses down to 2 ns, as well as an operation up to a switching frequency of 100 MHz. To control a PMOS high-side switch, a level shifter based on a symmetrical single-stage amplifier with minimized propagation delay is suitable. To control an NMOS high-side switch in a switching high-side supply rail, a level shifter with dedicated clamping structures is necessary to make the level shifter robust and insensitive to coupling during the converter's switching transitions. The clamping structures require to have minimized parasitic capacitances in the signal path, and limit the signal swing to achieve fast transitions. A proposed level shifter suitable for NMOS high-side switches, safely transfers on-time pulses of less than 3 ns with a propagation delay of below 5 ns to the high-side domain, which is confirmed by experimental results of a buck converter at switching transitions up to 20 V/ns; simulations even confirm the robustness up to 80 V/ns. A split-path gate driver with asymmetrically sized tapered buffers is suitable for minimum propagation delay, strongly driving pulses as short as 3 ns at the gate of the power switches. It achieves a propagation delay of typically 4 ns, and reduces the current consumption by up to 23%.

The main loss contributors at high input voltages and multi-MHz switching are parasitic capacitances at the switches and freewheeling diode in an asynchronous buck converter, and the dead time in addition in a synchronous buck converter. The non-linear voltage dependence of the most critical capacitances, for example of the freewheeling diode, needs to be modeled highly accurate to avoid large errors in the efficiency model. A piece wise linear model for transition losses (VI overlap) is required, as the influence of non-linear capacitances results in non-linear transitions, which are not covered sufficiently in conventionally used efficiency models. An efficiency model based on a four-phase model allows to separate loss causes and loss locations. A efficiency model for high input voltages and fast-switching converters proposed in this book matches efficiency measurements by less than 3%.

A dead time loss model allows to extensively analyze the influence of the dead time on efficiency in a synchronous buck converter. A predictive mixed-signal dead time control, which adjusts the dead time for the subsequent period, overcomes the propagation delay of the detection circuits, limiting the minimum possible dead time, and completely avoids body diode conduction and its related losses. A large dead time range even achieves soft-switching at low-side switch turn-on. The dead time detection is realized by sampling the switching node at the low-side switch turn-on and turn-off. An 8-bit timing control by differential delay lines allows to precisely control the dead time to a defined target window with a resolution of 125 ps over a dead time range of 32 ns. The values for both the resolution and the dead time were derived from the dead time model. Experimental results of the implemented dead time control confirmed a loss reduction of up to 30%, resulting in an efficiency improvement of 6% at 10 MHz switching and 12 V input. Measurements confirm a resolution of about 100 ps, which is the highest achieved dead-time resolution compared to state-of-the-art. An extension of the dead time control for light-load conditions achieves soft-switching at the low-side and high-side switch, and thus, is especially effective at loss critical high input voltages. Experimental results confirm a loss reduction of up to 45% for input voltages as high as 48 V, resulting in up to 14% efficiency improvement.

Resonant converters are investigated, which improve dominant switching losses by soft- or zero-voltage switching. Conventional quasi-resonant converters (QRC) require a large switching frequency range of about 10–50 MHz over a wide input voltage range and output current range. In addition, soft-switching is only achieved in a limited operating range. For this reason, a parallel-resonant converter (PRC) is proposed, which achieves soft-switching over a wide input voltage range of 12–48 V, and over an output current range of 100–500 mA. Supported by a fully integrated adjustable 5-bit capacitor array and configurable number of resonant oscillations, the PRC reduces the switching frequency range to 9–15 MHz, which is more than five times lower compared to conventional resonant converters, like the QRC, with the same input voltage and output current range. Experimental results confirm that the proposed PRC achieves peak efficiencies up to 76.3% at

10.8 MHz switching. Frequencies up to even 25 MHz are supported below input voltages of 24 V by changing the operating mode. Thus, the PRC operates at switching frequencies, which were not covered previously by state-of-the-art high-V_{in} converters with fully integrated power switches. A comparison using a proposed design indicator demonstrates that the PRC and the proposed dead time control of this book are especially effective at high input voltages. It further shows that different architectures result in a similar efficiency performance, while a change to an improved switch technology, from e.g., silicon to gallium nitride, enables a major improvement, independent of the converter architecture.

8.2 Comparison to State-of-the-Art

In Fig. 8.1, several operating points of the converters proposed in this book are compared to state-of-the-art converters presented in Sect. 2.4, Fig. 2.19a. The converters of this book cover a wide input voltage range up to 48 V. Figure 8.1a shows that the synchronous buck converter with dead time control and especially the parallel-resonant converter (PRC) reach power efficiencies close to the efficiency range of state-of-the-art commercial devices, while the switching frequencies of the converters of this book are up to 20 times higher at comparable input voltages. Figure 8.1b demonstrates that the converters of this book cover a frequency range for converters with input voltages above 5 V, which were not addressed by state-of-the-art converters previously. The converters of this book are closing the gap towards the

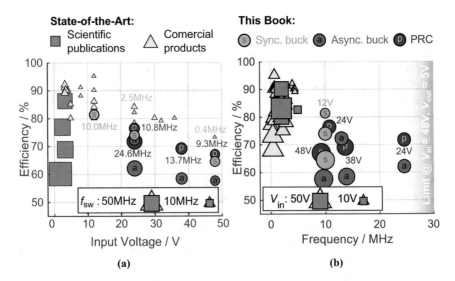

Fig. 8.1 Efficiency of published state-of-the-art converters and converters available on the market depicted over (**a**) the input voltage V_{in} and (**b**) the switching frequency f_{sw} for converters with $V_{in} > 5$ V

technology limitation of the switching frequency described in Sect. 2.3.6, Fig. 2.18a, which is reached at about 30 MHz for a conversion from 48 V to 5 V. The PRC and the asynchronous buck converter operate close to this frequency at 24 V input. The circuit blocks of the asynchronous and synchronous buck converters of this book have been proven to be able to operate up to at 48 V input at duty cycles which would allow switching frequencies above 30 MHz. However the expected efficiency would drop significantly below 70%, which is mostly not acceptable anymore in real applications.

It can be concluded that an implementation of converters with integrated power switches can be designed and operated up to the structural limit of the switching frequency. Advanced switching techniques like dead time control and soft-switching, as proposed in this book, significantly improve the efficiency. This enables a major increase of the switching frequency compared to state-of-the-art converters, and enables a significant size and cost reduction. However, a further increase of the switching frequency is still limited by power losses. In consequence, the efficiency most likely decreases below the acceptance level, before the structural limitation of the switching frequency is reached.

8.3 Outlook

This book presented various aspects, at which the available technology leads to a limitation for high-V_{in} multi-MHz converters. Furthermore, several aspects and ideas were identified, which were not covered and investigated in the scope of this book, but have a high potential to further improve these converters.

A major general limitation for the implementation of converter, especially with a floating NMOS high-side switch, is the charge coupling to the substrate and interfering with the control circuits. A back-side metalization was shown to be efficient, but it might not be suitable in real products due to an extra production step. Furthermore, it is only effective if the back-side is connected with a low inductance back to the top-side reference ground, if not, parasitic inductances of the connection generate again a ringing of the substrate. To overcome this risk, a substrate protection should be intrinsically available, which was not the case in the technology used for the converters of this book. Conducting deep-trenches have been shown in simulation to be the most suitable option. Also silicon-on- insulator (SoI) may be a promising technology. Alternatively, technologies are available, which use a low-doped, low-resistive substrate, which might be the better choice for switching converters, as the injected substrate noise is poorly propagated across the chip.

The proposed parallel-resonant converter (PRC) is shown to be up to about 15% more efficient than an asynchronous buck converter. However, its full potential in terms of efficiency could not be exploited, because the resonant power switch was implemented as a non-optimized standard drain-extended PMOS transistor. A PMOS switch had to be chosen, as the resonant note oscillates down to $-V_{in}$

in worst case. A suitable optimized NMOS switch was not present in the chosen technology, which is able to have significant negative voltages at its source. Efficiency estimations showed a potential improvement with an NMOS switch by up to 10%.

Several architectural changes could be applied to even further improve the efficiency of the PRC. It is shown in Figs. 7.14 and 8.1 that the PRC is up to nearly 15% higher in efficiency compared to the asynchronous buck converter. A comparison of the buck converter architectures showed that a synchronous converter with dead time control exceeds the efficiency of an asynchronous converter by up to 5%, or even up to 10% at high V_{in}. As the PRC is operated with an asynchronous output stage, the PRC would benefit from this improvement in efficiency if the output stage is changed to a synchronous output stage with the proposed dead time control. Efficiencies significantly above 80%, potentially reaching even 90% could be possible, even at input voltages up to 48 V at switching frequencies above 10 MHz. Soft-switching converters as proposed in this book might be especially beneficial at even higher input voltages above 50 V.

A further efficiency improvement is possible, by using external power switches within the proposed converters of this book, especially gallium-nitride (GaN) switches, enabling also higher output power. A major challenge is the required short connection of the gate driver to the external switches. New 3D assembly techniques as systems-in-a-package (SiP) [2], which allows a close integration of the switches to the control circuits, are essential. An even more advanced approach could be a full integration of the power stage, including level shifter and gate driver into the external power switch in a high-voltage technology. This is demonstrated in [1, 3– 5]. Implementing the high-voltage part and the power stage of the converter entirely in, e.g., GaN, would allow to use a low-voltage silicon technology to implement the low-voltage control circuits, while the silicon technology could be chosen to be optimized for bandwidth, rather than high-voltage capability. 3D integration would further allow to closely integrate the required buffer capacitors for the input voltage, and the low- and high-side gate driver supply, and thus further improve ringing and the switching behavior.

A future work also has to cover regulation and stability of high-V_{in} multi-MHz converters. A stable operation of the proposed converters was confirmed on silicon, however, a complete description of the regulation loop was not considered part of the scope of this book. It should be theoretically confirmed to have no interference of the dead time control loop to the main converter loop. A description of the regulation loops is in particular challenging for the PRC, as several regulation loops are interacting, which are the control of the main and the resonant switch, the on-times of both switches, as well the control of the integrated capacitor array. Rather than proving stability, it is to investigate, how the particular loop should be designed, to achieve an optimum transient performance of the output voltage, and also to regulate fast into soft-switching condition.

A comparison with the proposed design indicator in Fig. 7.16 indicated that hybrid converters could potentially achieve a similar efficiency performance in multi-MHz high-V_{in} operation if the circuit design and switching techniques proposed in this book are applied to, e.g., a three-level buck converter.

In this book, it is shown that the switching frequency relates to the volume of the inductors. However, the final size and volume of the converters are also dependent on the design of the inductors, which is only partially covered in this book. An effective reduction of the size due to the achieved increase of the switching frequency of the converters is only possible by including the inductor design in the overall implementation. This is in particular important for the design of the resonant inductor in the PRC, as commercial inductors are typically not optimized for the switching conditions and the current profile of the resonant inductor.

References

1. Han SW, Park SH, Kim HS, Jo MG, Cha HY (2017) Normally-off AlGaN/GaN-on-Si MOS-HFET with a monolithically integrated single-stage inverter as a gate driver. Electron Lett 53:198–199. https://doi.org/10.1049/el.2016.2813. https://ieeexplore.ieee.org/document/7843829/
2. Lau JH (2011) Evolution, challenge, and outlook of TSV, 3D IC integration and 3D silicon integration. In: 2011 International symposium on advanced packaging materials (APM), pp 462–488. https://doi.org/10.1109/ISAPM.2011.6105753
3. Mehrotra V, Arias A, Neft C, Bergman J, Urteaga M, Brar B (2016) GaN HEMT-based >1-GHz speed low-side gate driver and switch monolithic process for 865-MHz power conversion applications. IEEE J Emerg Sel Topics Power Electron 4:918–925. https://doi.org/10.1109/JESTPE.2016.2564946. https://ieeexplore.ieee.org/document/7469808/
4. Moench S, Kallfass I, Reiner R, Weiss B, Waltereit P, Quay R, Ambacher O (2016) Single-input GaN gate driver based on depletion-mode logic integrated with a 600V GaN-on-Si power transistor. In: 2016 IEEE 4th workshop on wide bandgap power devices and applications (WiPDA), pp 204–209. https://doi.org/10.1109/WiPDA.2016.7799938. https://ieeexplore.ieee.org/document/7799938/
5. Reiner R, Waltereit P, Weiss B, Moench S, Wespel M, Müller S, Quay R, Ambacher O (2018) Monolithically integrated power circuits in high-voltage GaN-on-Si heterojunction technology. IET Power Electron 11:681–688. https://doi.org/10.1049/iet-pel.2017.0397. https://ieeexplore.ieee.org/document/8338223/

Index

A

Automotive, 93
    48V battery, 11
    board net, 9
    electric utility vehicles, 11
    electric vehicle, 9
    fully-electrical turbocharger, 10
    hybrid vehicles, 11
    light-electric vehicles, 11
    starter-generator, 10

B

Back-side metalization, 60, 165
Bandgap reference, 129
Body diode, 107, 143, 163
    conduction, 151
Boundary conduction mode, 136
Buck
    current-mode control, 68
    minimum on-time, 39
    switching transition, 40
    voltage mode control, 67
Buck converter, 144
    asynchronous, 24, 47, 53, 78, 83, 89, 92,
        114, 117, 162, 166
    discontinuous conduction mode, 48
    hard-switching, 152
    multi-level, 26
    synchronous, 24, 47, 53, 89, 104, 114, 136,
        155, 162, 166
    three-level, 26, 167
    voltage regulation, 67

C

Capacitor
    array, 145, 163
    charging losses, 100
    non-linearity, 101
    poly-Si, 148
    scaling, 33
Cascaded converter, 17
CMOS technologie, 51
Continuous conduction mode (CCM), 48, 137
Counter
    cycle-by-cycle, 147
    up-down, 127, 147
Critical conduction mode, 136
Current sensor, 68

D

Data center, 12
    power usage effectiveness, 12
DC-DC converter
    point-of-load, 15, 161
Dead time, 47, 117, 125, 130
Dead time control, 24, 38, 109, 113, 117, 125,
        132, 133, 157, 162, 166
    digital, 126
    light load, 24, 110, 136, 163
    Maximum Power Point Tracking, 125
    predictive, 125
Deep trench, 162, 165
Delay chain, 147
Delay element, 130
Design indicator, 117, 156, 164, 167

© Springer Nature Switzerland AG 2020   169
J. Wittmann, *Integrated High-V_{in} Multi-MHz Converters*,
https://doi.org/10.1007/978-3-030-25257-1

Differential delay chain, 130
Differential delay line, 163
Discontinuous conduction mode (DCM), 136
Diverting structure, 58
Doping profile, 58
Drones, 11

E
E-bike, 11
Efficiency, 114, 117, 118, 135, 138, 141, 145,
        151, 155, 165
    scaling, 117, 118
Efficiency model, 89, 98, 103, 163
    four-phase model, 96
    loss causes, 98
    loss locations, 98
Electrical motor, 15
Electronic control unit, 9
E-mobility, 9
Epitaxial Layer, 57

F
Forced continuous conduction mode, 136
Freewheeling diode, 104, 137, 163
Full integration, 12

G
Gallium nitride, 52, 157
Gate driver, 37, 39, 49, 53, 67, 79, 84, 166
    asymmetry factor, 82
    cascaded, 82
    high-side supply, 76, 77, 166
    split-path, 162
    tapered buffer, 82, 162

H
Hard-switching, 22, 151
High-side isolation, 84
High-side supply, 116, 162
    bootstrap, 116
    charge pump, 116
    linear regulator, 116
Hybrid converter, 157, 167

I
IC package, 53
Inductance
    bond wire, 53
Inductor, 145, 161, 167

air core, 154
core losses, 30
resonant, 142, 145, 167
scaling, 30
Industry 4.0, 15
Integrator, 148
Internet of Things (IoT), 15
Isolation structure, 56, 57

L
Laterally diffused metal oxide semiconductor
        (LDMOS), 51
Level shifter, 39, 49, 67, 116, 162, 166
    capacitive, 77
    conventional, 76
    NMOS, 78, 84
        clamping structure, 81
        class AB, 81
        coupling currents, 81
        overlapping clamping, 79
    PMOS, 77, 84
        pre-bias current, 78
Linear regulator, 11, 18, 70
Logic, 148
Losses, 162
    body diode conduction, 90
    causes, 89, 163
    conduction, 35, 91, 107, 114, 136, 163
    control circuits, 92
    cross conduction, 107, 127, 138
    dead time, 163
    dead time related, 89, 105
    diode conduction, 93, 144
    ESR, 92
    gate charge, 145
    gate driver, 92, 104
    inductor, 35, 36, 91, 113, 155
    locations, 89, 163
    normalized, 118
    parasitic capacitances, 91, 104
    parasitic inductance, 92
    reverse recovery, 90, 107, 138, 144
    switching, 35, 106, 113, 114, 136, 154
    transition, 35, 90, 91, 96
    transition losses, 24

M
Maximum detection, 137, 147
Metal-oxide field-effect transistors (MOSFET),
        52
Micro-controller, 9
Miller amplifier, 129

Miller capacitance, 35, 90
Moat contacts, 58
Mobile devices, 16

N
N-guard ring, 60

O
Oscillation, 143
    frequency, 146
    multiple, 145
    period, 143, 152

P
Parallel-resonant converter (PRC), 78, 83, 138,
    144, 155, 163
PCB, 53, 132, 149
P-guard ring, 58, 60
Point-of-load converter, 39
Power efficiency, 12
Power stage, 49
    NMOS, 83, 145
    PMOS, 83, 145
Power supply unit, 13
Power switch, 39, 51
    high-side, 75, 78
    integrated, 40
    Miller plateau, 82
    NMOS, 78
Propagation delay, 76, 162
Propagation delay compensation, 131
PWM
    comparator, 67, 74
    generator, 39, 67, 69, 83, 148, 161, 162
    multiplexer, 72

Q
Quasi-resonant converter, 83

R
Ramp generator, 73
    integrator, 73
    symmetrical amplifier, 73
Range monitor, 147
Resonant circuit, 141, 145
Resonant converter, 22, 39, 141
    quasi-resonant, 22, 141, 163
Resonant switched-capacitor converter, 26
Reverse recovery, 38, 47, 107

S
Safety extra-low voltage (SELV), 15
Sample and hold, 128, 149
Sawtooth generator, 70, 71, 162
Schottky diode, 47, 89, 93, 114, 144, 155
    capacitance, 95
Server, 12, 161
    blade, 13
    power usage effectiveness, 13
    rack, 12, 14
Silicon technology, 40
Single stage conversion, 17
Soft-switching, 39, 101, 136, 139, 141, 151,
    157, 163, 166
State-of-the-art converter, 6, 37, 89, 135, 144,
    156, 161, 164
Substrate, 56, 165
    contacts, 57
    coupling, 116, 162
    isolation, 162
Switch control, 147
Switched-capacitor converter, 20
System-in-a-package, 166
Systems-on-a-chip (SoC), 1, 9, 16

T
TCAD simulation, 58
Technology
    gallium-nitride, 162
    high-voltage, 144, 166
    high-voltage BiCMOS, 149

U
USB, 16
    charger, 16

V
Voltage dividor, 149
Voltage regulator, 14

W
Window comparator, 129

Z
Zero-cross detection, 147
Zero-voltage switching (ZVS), 22, 101, 107,
    113, 127, 137, 138, 141, 145, 163
Zero-voltage switching quasi-square-wave
    converter (ZVS-QSW), 137